A Revolutionary War Road Trip on US Route 60

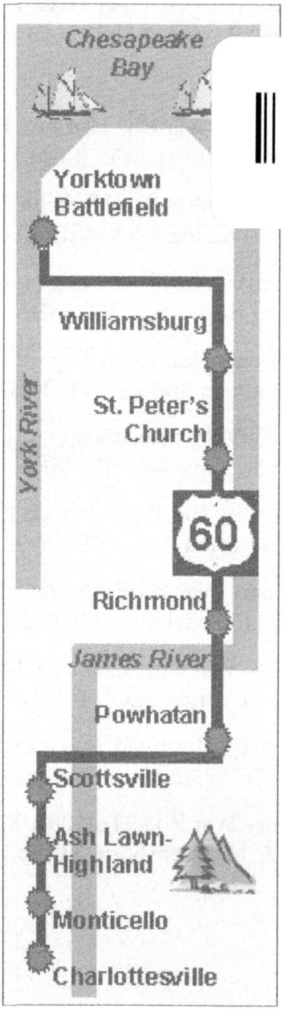

Spend a Revolutionary Day Along One of America's Most Historic Routes

OTHER "REVOLUTIONARY DAYS"
By Raymond C. Houghton

A Revolutionary War Road Trip on US Route 4
(Castleton, VT to Albany, NY, ISBN 1-931373-09-4)

A Revolutionary War Road Trip on NY Route 5
(Oswego to Schenectady, NY, ISBN 1-931373-21-3)

A Revolutionary War Road Trip on US Route 7
(Pittsfield, MA to Burlington, VT, ISBN 1-931373-10-8)

A Revolutionary War Road Trip on US Route 9
(Kings Ferry to Saratoga Springs, NY, ISBN 1-931373-12-4)

A Revolutionary War Road Trip on US Route 9W
(New York City to Kingston, NY, ISBN 1-931373-11-6)

A Revolutionary War Road Trip on US Route 20
(Pittsfield to Boston, MA, ISBN 1-931373-19-1)

A Revolutionary War Road Trip on US Route 202
(Elk Neck, MD to Philadelphia, PA, ISBN 1-931373-13-2)

A Revolutionary War Cruise on the Champlain Canal
(Bethlehem, NY to Vergennes, VT, ISBN 1-931373-14-0)

A Revolutionary War Road Trip on US Route 221
(Chesnee, SC to Augusta, GA, in preparation)

A Revolutionary War Road Trip on US Route 15-501
(Camden, SC to Greensboro, NC, in preparation)

A Revolutionary War Road Trip on US Route 60

Spend a Revolutionary Day Along One of America's Most Historic Routes

Raymond C. Houghton

Cyber Haus
Delmar, NY

A Revolutionary War Road Trip on US Route 60
Spend a Revolutionary Day Along
One of America's Most Historic Routes

Revolutionary War Road Trips
are published by Cyber Haus and
printed on-demand by
Amazon.com's Booksurge Division,
www.booksurge.com, 1-866-308-6235.

ISBN: 1-931373-20-5

Cyber Haus
159 Delaware Avenue, #145
Delmar, NY 12054
www.revolutionaryday.com
www.cyhaus.com
cyhaus@msn.com
518-478-9798

To my youngest daughter,
Amy, who endured many
long days with her parents
along historic Route 60.

INTRODUCTION

US Route 60 extends coast-to-coast, from Los Angeles, CA to Norfolk, VA, but in its last 150 miles, Route 60 tracks the end of the British Southern Campaign to its conclusion at the Battle of Yorktown, the last major battle of the Revolutionary War.

This campaign began with three invasions of Virginia by sea and ended with a fourth invasion from the south by land. During the four invasions, British forces ran roughshod over Virginia until the Americans surrounded them at Yorktown. General Charles Lord Cornwallis, the British commander of the southern campaign, saw Virginia as the keystone to the American rebellion. It was the largest colony and one of the richest. It was also centrally located, and control of the state would cut-off supplies and communications to the Americans.

General Sir Henry Clinton, commander of all British forces, began the campaign by sending the first invasion force in May of 1779 from his headquarters in British-held New York City. This army attacked and destroyed Virginia's eastern ports, including Norfolk, Portsmouth and Suffolk. Not far from this attack was Williamsburg, the colonial capital of Virginia. Later in 1780, the legislature decided to move the capital to Richmond, which was viewed as a safer, more central location.

Revolutionary War Events Along Today's US Route 60

1775

- Patrick Henry delivers his famous "Liberty or Death" speech in Richmond
- Royal Governor of Virginia, Lord Dunmore, flees Williamsburg
- Patrick Henry becomes the first Colonial governor of Virginia

1776

- The Second Continental Congress ratifies the Declaration of Independence drafted by Thomas Jefferson from Charlottesville
- James Monroe, also from Charlottesville, serves under George Washington at Trenton

1778

- British prisoners captured at the battles of Saratoga are moved from Boston to Charlottesville

1779

- Norfolk, Portsmouth and Suffolk destroyed by first British invasion of Virginia
- Thomas Jefferson becomes governor of Virginia

1780

- The capital of Virginia is moved from Williamsburg to Richmond
- Benedict Arnold leads second British invasion up the James River to Richmond

1781

- Phillips leads the third British invasion up the James River to Petersburg
- Cornwallis leads the fourth British invasion and takes charge of all British forces in Virginia
- British surrender at Yorktown

The second invasion sent by General Clinton arrived in December of 1780 and was led by the newly commissioned British General and American traitor, Benedict Arnold. As General Arnold's army began to threaten Richmond, the capital of Virginia, the Virginia legislature moved their proceedings to Charlottesville.

General Clinton's third invasion arrived in March of 1781 under the command of General William Phillips. He had been sent to take charge of all British forces in Virginia, as Clinton still did not completely trust the turncoat, Benedict Arnold. Phillips, who outranked Arnold, could keep an eye on him, but Phillips contracted typhoid fever. When the disease eventually took his life, General Arnold took charge of the British forces of the combined armies of the second and third invasions.

The fourth and largest invasion led by General Cornwallis began deep in the south with successes in South Carolina and Georgia. Despite the southern victories, the British were unable to completely destroy the Southern Department of the Continental Army and chased them northward toward Virginia. In May of 1781, General Cornwallis invaded Virginia and took command of all British forces in Virginia including General Arnold's combined second and third invasion forces. The campaign would end with the American siege at Yorktown and Cornwallis' surrender, effectively ending British attempts to regain control of the American colonies.

A Revolutionary Day along Historic US Route 60 begins early in the morning at the University of Virginia and downtown Charlottesville, where an early warning allowed the Governor of Virginia, Thomas Jefferson, and most of the Virginia Assembly to escape British capture.

From Charlottesville, the road trip heads south to the James River with stops at Monticello, the home of Thomas Jefferson and Ash Lawn-Highland, the home of President and Revolutionary War Officer, James Monroe.

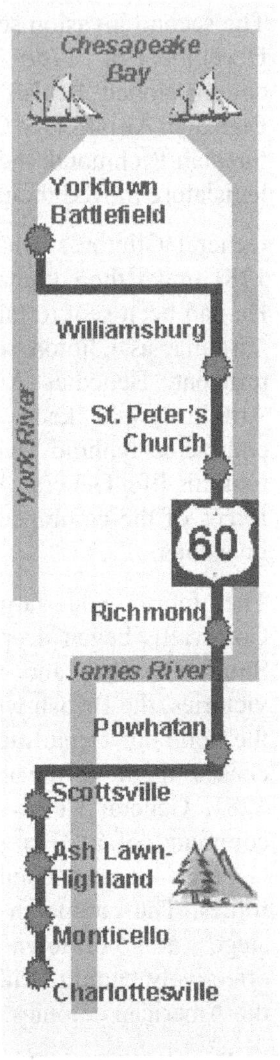

After crossing the James River at Scottsville, the road trip joins US Route 60 and turns east through Powhatan to Richmond, the capital of Virginia from 1780 to present. Before Richmond became the capital, it hosted the Second Virginia Convention in March 1775. During the convention, Patrick Henry delivered his famous "Liberty or Death" speech, which echoed throughout the thirteen colonies and became the clarion call to revolution.

From Richmond, the road trip continues east through New Kent to Williamsburg, the second capital of Virginia and center of Revolutionary War activities for most of the war. Today, Colonial Williamsburg is one of the oldest and largest living history museums.

From Williamsburg, the road trip continues east along the Colonial Parkway to the Yorktown battlefield. In October 1781, the last major battle of the Revolutionary War would be fought here.

So, if you're ready, begin your **Revolutionary War Road Trip on US Route 60.**

TABLE OF CONTENTS

AND

TRIP LOG

ACKNOWLEDGEMENTS

I would like to thank those who provided feedback and editing support on earlier formats of this book, especially my daughter, April Monica Houghton Blakeslee; my friend and mentor, Russ Denegar; and those who posted feedback on the Internet from the Route 60 web site. Finally, I would like to thank my wife, Jan Laws Houghton, for her loving support and encouragement.

UNIVERSITY OF VIRGINIA
CHARLOTTESVILLE
VIRGINIA

The University of Virginia stands on land purchased in 1788 by a Revolutionary War veteran, James Monroe, who would become the fifth President of the United States thirty years later. As an eighteen-year-old Lieutenant, Monroe served with George Washington at the Battle of Trenton and was wounded while leading a charge that would help achieve a crucial victory for General George Washington.

Later, while Monroe served as President, Thomas Jefferson, the author of the Declaration of Independence and Governor of Virginia during the four British invasions of Virginia, turned

James Monroe's First Farm
Site of the University of Virginia

In 1788 James Monroe purchased an 800-acre farm here to be close to his friend Thomas Jefferson and to establish a law office. In 1799 the Monroes moved to their new Highland plantation adjacent to Monticello and sold the first farm. In 1817 the Board of Visitors of Central College purchased 43 3/4 acres of Monroe's old farm, for the Lawn and the Ranges of the "academical village" that Jefferson was planning to build with private contributions. On 6 October, President Monroe, with former presidents Jefferson and Madison, laid the cornerstone for its first building, Pavilion VII. On 25 January 1819, Central College was charted by the General Assembly as the University of Virginia.

(UVA Marker)

Monroe's farmland into an "academical village." After Jefferson completed his term as the third president of the United States, he began construction of the University of Virginia in 1817. He surveyed the site himself using a theodolite, locust stakes and twine. The University was founded by the state in 1819 and the "Academical Village" was completed, except for the Rotunda, by 1823. Classes began in 1825.

In creating the University, Jefferson brought together his interests and talents in architecture, education and gardening. A walk around the old campus reveals his design, which included a rectangular, terraced green space called the Lawn, two parallel rows of buildings,

Cabell Hall

Dedicated 1898, Stanford White, Architect.

Named in honor of Joseph Carrington Cabell without whose aid the university might never have come into being. He later served as rector succeeding Thomas Jefferson and James Madison.

And you will know the truth, and the truth will make you free. (Translation of the Greek inscription on the frieze)

This tablet given by the Seven Society.

(UVA Marker)

pavilions, colonnaded walk-ways and the Rotunda.

His innovative model kept students and faculty close. They lived, worked, dined and studied within steps of each other. He hoped that the closeness would spark the intellectual life of the academic community.

Today, the University has grown significantly with almost 20,000 students, ten degree-granting schools and about 2,000 full-time faculty members. However, despite this increase in size, a visit to the "Academical Village," the old campus and center of the University, makes it clear that Jefferson's vision of student and faculty closeness still remains the focus of the University.

UNIVERSITY OF VIRGINIA TO DOWNTOWN CHARLOTTESVILLE

Mile Mark 0.0 — Depart the University of Virginia on University Avenue heading towards downtown Charlottesville.

Mile Mark 0.2 — Watch for the College Inn on the left. The inn is a great local restaurant and is a recommended stop for breakfast.

Mile Mark 1.0 — Reach the Lewis and Clark statue in the center of an intersection with McIntire Road. Meriwether Lewis, who was born during the time of the Revolutionary War in the Charlottesville area, was private secretary to President Thomas Jefferson in 1801. Lewis and Clark were both interested in exploring a land route to the Pacific Ocean. In 1804, they began their expedition.

Turn left onto McIntire Road. Along the way, you'll pass a sign with the silhouettes of the three presidents from the Charlottesville area: Madison, Jefferson and Monroe. Jefferson and Monroe's homes are upcoming sites on this road trip. Madison's home is 27 miles north of Charlottesville in Montpelier Station.

Madison was the fourth president of the United States and, like Monroe, he was a close friend to Thomas Jefferson. During the Revolutionary War, Madison served in the Continental Congress. Later, he would serve in the Constitutional Convention and became known as "the Father of the Constitution."

Mile Mark 1.3 — Reach an intersection with Preston Avenue. Turn right and make an immediate left onto High Street.

Mile Mark 1.6 — Climb the hill and watch for the northwest corner of Court Square on the right. The square is your first destination in downtown Charlottesville.

Mile Mark 1.7 — Reach the intersection with Park Street. Turn right on Park Street and find a parking place for a visit to Court Square.

Arrive in downtown Charlottesville.

Charlottesville

The site was patented by William Taylor in 1737. The town was established by law in 1762, and was named for Queen Charlotte, wife of George III. Burgoyne's army, captured at Saratoga in 1777, was long quartered here. The Legislature was in session here, in June 1781, but retired westward to escape Tarleton's raid on the town. Jefferson, who lived at Monticello, founded the University of Virginia in 1819.

Conservation & Development Commission 1929

(Charlottesville Marker)

CHARLOTTESVILLE
VIRGINIA

Charlottesville was founded in 1762 on Three Chopt Road, a major east-west thoroughfare.

The Revolutionary War first came to Charlottesville in 1778 when the prisoners captured during the battles of Saratoga in the fall of 1777 were moved from Boston to Charlottesville.

Saratoga was a major American victory, which was, in large part, due to the heroic actions by then American General Benedict Arnold, who was wounded during the battle. About 6,000 soldiers from British General John Burgoyne's forces surrendered their weapons at Saratoga. The victory has been called the "turning

Courthouse

First portion built between 1763 and 1781. Added to in 1803 and 1860. Remodeled and restored, 1938.

(Charlottesville Marker)

point" of the Revolutionary War because it helped to convince the French to aid the American cause.

In Charlottesville, the prisoners were required to build their own housing and be present for roll calls. They did, however, enjoy some degree of freedom, such as holding jobs in town. Officers could arrange their own housing, so long as it was within a day's ride from camp.

As the second British invasion of Virginia, led by now British General Benedict Arnold, began to threaten Richmond in December of 1780, the Virginia legislature moved their proceedings to Charlottesville. After General Cornwallis had taken charge of all British forces in Virginia in early June of 1781, he sent Colonel Banastre Tarleton and a mounted British force on a lightning raid to Charlottesville. He hoped to capture members of the Virginia legislature and Governor Thomas Jefferson. Unfortunately for Tarleton, Captain John Jouett of the Virginia Militia galloped ahead of Tarleton and was able to spread the alarm.

Site of Old Swan Tavern where lived and died Jack Jouett whose heroic ride saved Mr. Jefferson, the governor, and the Virginia assembly from capture by Tarleton, June 1781.

Erected by the Monticello Branch of the Association for the Preservation of Virginia Antiquities, 1910.

(Charlottesville Marker)

Jack Jouett's Ride

On 4 June 1781, John "Jack" Jouett Jr. arrived at the Albemarle County Courthouse to warn the Virginia legislature of approaching British troops. The state government under Governor Thomas Jefferson had retreated from Richmond to reconvene in Charlottesville because of the threat of British invasion during the Revolutionary War. Jouett had spotted Colonel Banastre Tarleton and his 180 dragoons and 70 cavalrymen 40 miles east at Cuckoo Tavern, and rode through the night to reach here by dawn. Jouett's heroic ride, which allowed Jefferson and all but seven of the legislators to escape, was later recognized by the Virginia General Assembly, which awarded him a sword and a pair of pistols.

Department of Historic Resources, 1998.

(Charlottesville Marker)

The Inn at Court Square

Built 1786.

(Charlottesville Marker)

After a brief skirmish at the edge of town, Tarleton entered a largely abandoned Charlottesville. Most of the legislature had escaped. Among those who didn't was the American Pioneer, Daniel Boone, who spent the night as a prisoner in a coal cellar. Boone and the other captured delegates were later paroled. For the most part, the Tarleton raid was not successful and he was only able to seize military supplies and tobacco.

Today, there are two markers at the Albemarle County Courthouse noting Jouett's ride. One is on the grounds of the courthouse and the other is across the street at the site of the Old Swan Tavern where Jouett lived and died.

President Monroe's Local Home

In 1788 James Monroe moved to Charlottesville and for one year his home was located in the first block west of this site. Then he lived for nine years in the home he built on what is now called "Monroe Hill" at the University of Virginia. His final Albemarle Home, near "Monticello" was his "Highland" estate now called "Ash Lawn."

(Charlottesville Marker)

Stone Tavern and Central Hotel

George Nicholas, Albemarle County's Virginia General Assembly delegate in 1783, built a stone house here in 1784. James Monroe occupied it 1789-1790, while improving the dwelling at his nearby farm, later the site of the University of Virginia. Here on 15 December 1806, while the house was being operated as the Stone Tavern, the return of Lewis and Clark was celebrated with a dinner. Thomas Jefferson hosted a reception in the tavern (renamed the Central Hotel), on 12 November 1824, for the Marquis de Lafayette. While serving as a hospital during the Civil War, the building burned with no fatalities in 1862.

(Charlottesville Marker)

Two other markers relating to the Charlottesville home of James Monroe can be found near the downtown mall at the corner of 5th and East Market Streets. The marker can be reached by car from the front of the courthouse by turning left on 4th Street and then the next left on East Market Street (4th Street is not well marked but is a one-way street heading south). The markers are on the right near 5th Street but are difficult to see.

CHARLOTTESVILLE TO MONTICELLO

Mile Mark 1.9 — Depart Charlottesville on East Market Street.

Mile Mark 2.0 — Reach 9th Street and the intersection with Route 20. Turn right on Route 20 heading south.

Mile Mark 2.4 — Turn left onto Monticello Avenue and continue to follow Route 20.

Mile Mark 2.8 — Pass a Charlottesville marker on the left.

Mile Mark 3.5 — Go under Interstate 64.

Mile Mark 3.9 — Move to the right lane as you approach the light at College Drive. Turn right at the light, then right again to the Monticello Visitor Center.

The exhibits at the visitor center are a great way to prepare for a visit to Monticello.

Monticello

Three miles to the southeast, Thomas Jefferson began the house in 1770 and finished it in 1802. He brought his bride to it in 1772. Lafayette visited it in 1825. Jefferson spent his last years there and died there, July 4, 1826. His tomb is there. The place was raided by British cavalry, June 4, 1781.

Conservation & Development Commission 1928

(Monticello Marker)

Mile Mark 4.2 — From the visitor center, return to Route 20 heading east, then immediately get into the left lane and turn left on Route 53 East.

Citizens of central and western Virginia have contributed significantly to national defense and to the US Army throughout its 200-year history. During the Revolutionary War, Virginians fought valiantly as members of the militia and Continental Army under Washington. Noteworthy is Captain Jack Jouett's all-night ride to Charlottesville in 1781 to warn Jefferson of the impending arrival of British cavalry. John Peter Muhlenberg, Lutheran clergyman in the Shenandoah Valley, became a general in the Continental Army and later served in Congress. George Rogers Clark, soldier and explorer, was instrumental in bringing the territory of six states into our nation. Its independence established, the nation expanded westward. Captain Meriwether Lewis and Lieutenant William Clark, both from Albemarle County families, explored the area of the Louisiana Purchase during their heroic expedition of 1804-06 ... All of these soldiers and many more, have established a proud tradition of service that continues this day.

(Visitor Center Marker)

The first schoolhouse Jefferson attended at Tuckahoe.

(Visitor Center Exhibit)

Mile Mark 5.2 — Watch for Mitchie Tavern on the right.

William Mitchie, a Revolutionary War veteran, was a corporal at Valley Forge when he was urgently requested to return home because of his ailing father. When he arrived, his father had passed away.

Mitchie inherited his father's land and established a tavern in 1784.

Today, visitors can experience the Tavern's past through exhibits, activities and a hearty midday fare offered by servers in period attire.

Mile Mark 5.9 — Bear right onto the entrance road to Monticello.

Mile Mark 6.4 — Reach the parking area and the Monticello Shuttle Station.

For your visit to Monticello, it is recommended that you take the shuttle to the house and then walk from the house down Mulberry Row to the cemetery, then take the cemetery pathway back down to the parking lot.

Arrive at Monticello.

MONTICELLO
VIRGINIA

Jefferson began constructing his Monticello, "little mountain," home in 1768. About the same time, he became a member of the Virginia House of Burgesses, which met in Williamsburg (an upcoming site on this road trip).

At the beginning of the Revolutionary War in 1775, Jefferson was a member of the Second Continental Congress in Philadelphia. It was there that he was drafted to write the Declaration of Independence, which was ratified with many revisions on July 4, 1776.

In 1779, Jefferson became Governor of Virginia. A year later, the capital of Virginia was moved closer to his Monticello home from Williamsburg to Richmond (also an upcoming site on this road trip).

While serving as Governor during the British second invasion of Virginia in January 1781, Jefferson

Thomas Jefferson's Monticello

Monticello, meaning "little mountain," was designed and built by Jefferson over a period of 40 years. One of our country's foremost architectural masterpieces, Monticello remains a testimony to its creator's ingenuity and diverse interests. The main house and gardens reflect the innovative character of their designer, while Mulberry Row offers insights into plantation industries and workers, both slave and free.

(Monticello Marker)

received a letter from Benedict Arnold offering to spare Richmond if he was allowed to seize tobacco unmolested. Jefferson refused and Arnold ransacked Richmond.

On June 4, 1781, Jefferson was entertaining the Speaker of the Virginia Assembly when he was warned about Tarleton's raid in Charlottesville. Jefferson made his escape by horseback barely minutes before Tarleton's arrival.

South Pavilion

Jefferson moved to Monticello in November 1770 taking up residence in this building, the first completed structure in his ambitious design for the mountaintop. Thirteen months later he brought his new wife, Martha, to live in this single room, with a kitchen on the lower level. They occupied this building while the main house was under construction.

The South Pavilion was later used as a schoolroom. In 1808, it was remodeled and two windows and a door were added.

Jefferson then housed his law books in this room, which was used as an office by young men studying law under his guidance. He invited a grandson-in-law to "join us at Monticello" and "to take his station among my law books in the South Pavilion."

(Monticello Marker)

Near the end of the war, Jefferson's wife, Martha, passed away. Two years later, Jefferson began service as trade commissioner and minister to France. He returned in 1789 with new ideas about architecture and he used these ideas to redesign Monticello. The redesign, with European influence, continued throughout his service as the third president of the United States and was largely completed when he returned from public service in 1809. Jefferson died here at Monticello on July 4, 1826, exactly fifty years after the signing of the Declaration of Independence.

Today, Monticello is a magnificent legacy to Jefferson. Led by a tour guide, visitors begin their tour at its entrance hall, an impressive two-story room that also served as Jefferson's museum. It is crowded with a wide variety of objects that he collected from all over the world. From the entrance hall, visitors are led through his library, bedroom, parlor, dining room and guest bedrooms. During the tour, visitors gain a great appreciation for the many talents and service of Thomas Jefferson.

After the tour, a walk down Mulberry Row is recommended. Mulberry Row was "main street"

Monticello. Along this street were workmen's homes, gardens, a smokehouse/dairy, a blacksmith shop/nailery, a joinery, a carpenter's shop and sawpit. Mulberry Row is evidence that Monticello was not only home to Thomas Jefferson but was also a community employing over a hundred slave/artisans.

Monticello and the Levy Family

This is the grave of Rachel Phillips Levy (1769-1839), daughter of Jonas and Rebecca Machado Phillips of Philadelphia, and mother of Commodore Uriah P. Levy, USN (1792-1862), who purchased Monticello in 1836. An ardent admirer of Thomas Jefferson, Commodore Levy believed that the houses of great men should be preserved as "monuments to their glory," and he bequeathed Monticello in his will to the "People of the United States." The government relinquished its claim to the estate, however, and litigation over the will deprived Monticello for seventeen years of an owner to care for it.

In 1879 Jefferson Monroe Levy (1852-1924), who shared his uncle Uriah's admiration for Jefferson, gained clear title to Monticello and began to make badly needed repairs. After adding considerable land from the original Monticello tract, he sold the house and 662 acres to the Thomas Jefferson Memorial Foundation in 1923.

At two crucial periods in the history of Monticello, the preservation efforts and stewardship of Uriah P. and Jefferson M. Levy successfully maintained the property for future generations.

(Monticello Marker)

Mulberry Row

This 1,000-foot-long road was the center of plantation activity at Monticello from the 1770's until Jefferson's death in 1826. It once was lined with the dependencies necessary to support the main house and a 5,000-acre plantation with as many as 150 residents. In 1796, there were seventeen structures on Mulberry Row, including dwellings for white and black workmen, wood and ironworking shops, a smokehouse-dairy, a washhouse, storehouses, and a stable.

(Monticello Marker)

Joinery

Remains of the joinery, or woodworking shop, 57' x 18'. Constructed of wood on a high stone foundation. Active from the 1770s to 1826.

(Monticello Marker)

The Monticello Labor Force

The operation of Monticello depended primarily on the labor of black slaves. In 1796, there were 110 African-Americans living on the 5,000-acre plantation, almost half of them children. Some of the slave-artisans and house-servants lived in five small log cabins on Mulberry Row. While these cabins have long since vanished, their foundations, cellars, and the discarded possessions of their residents have been uncovered by archaeologists.

Jefferson, who considered slavery an "abominable crime," acquired most of his slaves by inheritance. He freed only two in his lifetime and five in his will. The others were sold after his death to help pay the debts of his estate.

(Monticello Marker)

Life on Mulberry Row

Both black and white laborers lived and worked on Mulberry Row. Jefferson hired free white artisans to train his slaves, many of whom became highly skilled blacksmiths, carpenters, coopers, masons, charcoalburners, shoemakers, and weavers.

Jefferson provided each black family with a small dwelling, weekly rations of corn and pork, and summer and winter clothing. After their long work day (usually sunrise to sunset) and on Sundays and holidays, the Monticello slaves often supplemented their basic allotments by hunting, fishing, tending their gardens and poultry yards, or performing additional paid tasks.

(Monticello Marker)

Workman's House

Stone dwelling, 34' x 17' with stone chimney and brick floor. Constructed c1776 for white joiners and masons working on the Monticello house. Additions are modern.

(Monticello Marker)

Slave Cabin

Site of a dwelling, 20 1/2' x 12' - one of five slave cabins on Mulberry Row. Constructed of logs on a stone foundation, with wood chimney and earth floor. Occupied from the 1770s to c1800.

(Monticello Marker)

Although Jefferson called slavery "an abominable crime," he was trapped by the economics of the institution and was a slaveholder his entire life.

At the end of Mulberry Row is the family graveyard. The graveyard was laid out by Jefferson prior to the Revolutionary War and continues to be a family burying ground today. Over Jefferson's grave is a monument inscribed as he had instructed noting the three accomplishments of which he was most proud: (1) Author of the Declaration of Independence, (2) Author of the Statute of Virginia for Religious Freedom and (3) Father of the University of Virginia.

Here was buried Thomas Jefferson author of the Declaration of American Independence, of the Statute of Virginia for Religious Freedom, and Father of the University of Virginia. Born April 2, 1743 O.S. Died July 4, 1826.

(Monticello Marker)

The Monticello Graveyard

This graveyard had its beginning in an agreement between two young men, Thomas Jefferson and Dabney Carr, who were schoolmates and friends. They agreed that they would be buried under a great oak which stood here.

Carr, who married Jefferson's sister, died in 1773. His was the first grave on this site, which Jefferson laid out as a family burying ground. Jefferson was buried here in 1826.

The present monument is not the original, designed by Jefferson, but a larger one erected by the United States in 1883. Its base covers the graves of Jefferson, his wife, his two daughters, and Governor Thomas Mann Randolph, his son-in-law.

The graveyard remains the property of Jefferson's descendants, and continues to be a family burying ground.

(Monticello Marker)

MONTICELLO TO ASH LAWN-HIGHLAND

Mile Mark 6.4 — After returning to the parking area by the path from the Monticello Graveyard, depart Monticello.

Mile Mark 6.6 — Reach an intersection with Route 53. Turn left, following the signs to Ash Lawn-Highland.

Mile Mark 7.7 — Bear right and continue to follow Route 53 east. Pass the Jefferson Vineyards on the right.

Mile Mark 7.8 — Watch for the Colle marker, also on the right.

Colle is the home where the Saratoga prisoner, General Baron von Riedesel, and his wife, Baroness Riedesel, were held after the British prisoners of Saratoga were moved from Boston to

Colle

The house was built about 1770 by workmen engaged in building Monticello. Mazzie, an Italian, lived here for some years adapting grape culture to Virginia. Baron von Riedesel, captured at Saratoga in 1777, lived here with his family, 1779-1780. Scenes in Ford's novel, **Janice Meredith**, are laid here.

Conservation & Development Commission 1928

Charlottesville. General von Riedesel led the prisoner march and dined with Lafayette on the way. Not used to the southern climate, he suffered heat stroke and later was made a part of a British-American prisoner exchange. He returned to duty in British-held New York City in late 1779.

Mile Mark 7.9 — Bear right on Route 795, still following the signs to Ash Lawn-Highland.

Mile Mark 8.7 — Reach the entrance to Ash Lawn-Highland. Turn right.

Mile Mark 9.2 — Reach the parking area and visitor center.

ASH LAWN-HIGHLAND
VIRGINIA

Ash Lawn-Highland became the home of James Monroe about fifteen years after the end of the Revolutionary War. He moved here to be closer to his friend and mentor, Thomas Jefferson. The home, which originally faced Monticello, was remodeled by a later owner and now faces a courtyard behind a two-story addition.

James Monroe's Ash Lawn-Highland

A 535-acre estate, Ash Lawn-Highland was the Monroes' primary residence from 1799 until 1823. The original Monroe house features tours and exhibits on early nineteenth-century culture. Plantation life is explored through the original smokehouse and overseer's house, at the reconstructed slave quarters, and in the fields, where cattle and sheep graze and peacocks roam.

(Ash Lawn-Highland Marker)

George Washington Crossing the Delaware River

(Ash Lawn-Highland Exhibit)

Today, the home is owned and operated by the College of William and Mary (an upcoming site on this road trip). Both Jefferson and Monroe studied at the college prior to the Revolutionary War.

James Monroe

1758 - 1831

(Ash Lawn-Highland Marker)

Three Friends, Three Presidents

Three of the first five Presidents of the United States made their homes in central Virginia. Thomas Jefferson's Monticello and James Monroe's Ash Lawn-Highland are just outside Charlottesville. James Madison's Montpelier stands in neighboring Orange County. During Jefferson's Presidency, Madison served as Secretary of State while Monroe negotiated the Louisiana Purchase as Minister to France. Under Madison, Monroe served as Secretary of State and Secretary of War. Like their political lives, these men's social and personal lives were intertwined. Jefferson hoped to surround himself with his friends in central Virginia. Madison and Monroe were the core of Jefferson's "society to our taste."

(Ash Lawn-Highland Marker)

Tours of the home are led by college staff members. The tour starts with a discussion of the famous painting of Washington crossing the Delaware by Emanuel Gottlieb Leutze. Like many artists, Leutze took liberties in the portrayal of events in the painting. He painted Washington and Monroe standing in the boat with Monroe clutching a "Betsy Ross" flag, which at the time did not yet exist. Also notably missing from the painting is the driving snowstorm that took place the evening of the crossing.

Monroe, of course, did participate in the crossing of the Delaware and afterward, led a charge against the Hessian troops stationed in Trenton. Besides surviving wounds received during the charge, Monroe also wintered at Valley Forge. After the war, Monroe served Virginia and the United States in several capacities, including US Senator, foreign minister, Secretary of State during the War of 1812 under President James Madison and fifth President of the United States from 1817-1825.

Today, many of the objects that James Monroe collected during his public service, especially while he was a Minister of France, England and Spain, can be found on display in his Ash Lawn-Highland home.

ASH LAWN-HIGHLAND TO SCOTTSVILLE

Mile Mark 9.2 — Depart Ash Lawn-Highland.

Mile Mark 9.6 — Return back to the intersection with Route 795. Turn right and prepare for a drive through the scenic Virginia countryside along narrow and winding roads.

Mile Mark 11.4 — Reach an intersection with Rolling Road. Continue going straight across the intersection onto Route 627, although the route marker is small and difficult to see.

Mile Mark 13.0 — Pass the Lenark farm on the right.

Mile Mark 13.5 — Pass the Kluge Estate on the left.

Mile Mark 16.8 — Reach an intersection with Route 20. Turn left and rejoin Route 20 heading east.

Mile Mark 18.7 — Note that Route 20 is also called the "Constitution Route," in honor of James Madison.

Mile Mark 20.0 — Reach the Green Mountain Country Store - a great stop for coffee.

Mile Mark 24.1 — Pass the Fairview Mansion on the left.

Mile Mark 26.0 — Reach downtown Scottsville.

SCOTTSVILLE
VIRGINIA

Scottsville, which is on the James River, was a river port. Known as Scott's Landing during the war, it served as an important supply source for goods, such as tobacco. These supplies were floated down the river to Richmond in flat-bottom boats known as batteau.

Today, Scottsville is a scenic country town and a walk around town is recommended. There is an historic marker at the town's main intersection of Route 20 and Main Street. A short distance east of the marker is the Scottsville Museum at 290 Main Street. The museum includes exhibits from Scottsville's history

> **Historic Scottsville**
>
> In 1745, Old Albemarle County was organized at Scott's Landing, its first county seat, here on the Great Horseshoe Bend of the James River. In 1818, the town was incorporated as Scottsville. Beginning in 1840, it flourished as the chief port above Richmond for freight and passenger boats on the James River and Kanawha Canal. It played a vital role in the opening up of the west. The 1840's and 50's were its golden era.
>
> Virginia Historic Landmarks Commission 1976.
>
> (Scottsville Marker)

Tobacco was the mainstay of colonial Virginia's economy but soon depleted the rich agricultural lands of the coastal Tidewater region. As tobacco cultivation moved westward into the Piedmont, transporting the crop to market became increasingly difficult. Hauling hogsheads of tobacco overland was slow and expensive. The fastest route was by water. Dugout log canoes could navigate the James River's rapids, shallows and falls, but were too small and unstable for heavy cargo. To solve this problem, two long dugout canoes were tied together, side by side, to create a steadier craft that could carry greater loads. By the 1750s double canoes were a common sight on the James River as far west as Buchanan.

In 1771 Virginia's tobacco trade faced a crisis. A flash flood destroyed most of the dugout canoes used to carry crops to market. Replacing the lost boats was vital. However, the large trees needed for making the dugouts were no longer plentiful near the river. Brothers Anthony and Benjamin Rucker developed a new type of vessel, a flat-bottomed batteau that could be quickly constructed of planks, rather than whole trees. Designed for navigating the rocky river, this *James River Batteau* could carry larger loads and soon replaced the dugout as the chief freight boat on the upper James.

(Scottsville Exhibit)

with emphasis on its heyday as a port in the mid-1800's.

Across the street from the museum is an outdoor exhibit that also emphasizes Scottsville's history as a port on the James River. The exhibit covers the expansion of trade between Scottsville and Richmond and the development of the Kanawha Canal between the two areas.

St. John's Church

St. Anne's Episcopal Parish, Est. 1745.

(Church Marker)

Reversing direction on Main Street, just a block past Route 20, is Harrison Street. A short distance up Harrison Street is St. John's Church, which was established in 1745.

Lunch in Scottsville — There are several possibilities for lunch along Route 20 in Scottsville. Wherever you eat, it is likely that at one time the location was under water. Note the flood markers that can be found on many of the buildings downtown.

SCOTTSVILLE TO POWHATAN

Mile Mark 26.0 — Depart downtown Scottsville heading east on Route 20.

Mile Mark 26.2 — Cross the James River.

Cumberland County Courthouse

In 1749, the Virginia House of Burgesses divided Goochland County to establish Cumberland County. William A. Howard, an associate of Thomas Jefferson's master builder, Dabney Crosby, built the present Cumberland County courthouse (1818-1821). The unusual temple-form, Jeffersonian-Classical building has a finely executed Tuscan portico on the long side instead of the end and stands only one story high. Howard also designed the diminutive brick clerk's office to the east that was completed in 1821. Today, the two buildings stand near the center of the village with a jail (ca. 1823), a 19th-century well and a 1901 Civil War monument.

(Cumberland Marker)

Mile Mark 44.8 — Reach the intersection with US Route 15. Turn right for a brief jog over to US Route 60.

Mile Mark 46.2 — Reach the intersection with Route 60. Turn left and join Route 60 heading east.

Mile Mark 58.2 — Note the "Campaign of 1781'" marker on the right.

George Washington sent General Lafayette to Virginia with about 1,200 troops in response to the second and third invasions of Virginia. He wanted to stop Benedict Arnold's devastating raids on the towns and villages along the James River and possibly capture him. However, when Arnold's army linked up with Cornwallis, creating an army of about 7,000, Lafayette

Campaign of 1781

Steuben, both on his retreat from Simcoe and on his return north to join Lafayette, passed near here, June 1781.

(Route 60 Marker)

was greatly outnumbered. He wrote Washington: "It now appears that I have business with two armies."

Lafayette knew he had to keep his army mobile and could only skirmish with this large British army until he received reinforcements. To protect his supplies, he sent them west with General von Steuben to a site near this marker, a location Lafayette considered remote. However, British Colonel Simcoe with a small force learned of von Steuben's presence in the area. Simcoe tricked von Steuben by deploying his troops to look and sound like the entire vanguard of the combined British armies. Von Steuben retreated with his men, leaving the supplies in Simcoe's cunning hand.

Mile Mark 60.7 — Reach the Cumberland County Courthouse on the left. Note the historic marker in front of the courthouse.

Mile Mark 78.2 — Reach the Giles home marker on the right.

William Giles was thirteen at the beginning of the Revolutionary War. He pursued classical studies and attended the College of New Jersey (Princeton) during the war, graduating in 1781 at the age of 19. As noted

Giles's Home

Five miles southwest is the Wigwam, the home of William B. Giles, Jefferson's Chief Lieutenant; United States Senator, 1804-1815, and Governor of Virginia, 1827-1830, an orator and famous political leader. Giles died there, December 4, 1830

Conservation & Development Commission, 1934.

(Powhatan Marker)

by the marker, he later served with Thomas Jefferson and eventually became governor of Virginia.

Turn right after the Giles marker onto Route 1002.

Mile Mark 78.6 — Turn left onto Route 13 heading east for a visit to the town of Powhatan.

Mile Mark 80.2 — Reach downtown Powhatan.

POWHATAN
VIRGINIA

During the Revolutionary War, the town was known as Scottville (not to be confused with Scottsville, the previous site on this road trip). The town was later

Powhatan Courthouse

The first courthouse was built here about 1783 and around it grew the village of Scottville. Named for the Revolutionary War hero, General Charles Scott, who was born in the area. The town eventually became known as Powhatan Court House. The present courthouse was designed by Alexander Jackson Davis, of New York, and built in 1840. It is a masterpiece of Greek revival architecture. The oldest building in the court square is the clerk's office built about 1798 and nearby is a late-18th-century double-galleried courthouse tavern. Part of the Confederate army retreating from Richmond, passed here on 4 April 1865.

(Powhatan Marker)

named after the great Indian chief, Powhatan, father of the famous Indian princess, Pocahontas.

As noted on the courthouse marker, General Charles Scott was born in this area. General Scott served throughout the Revolutionary War.

At the Battle of Monmouth Courthouse in New Jersey in June 1778, General Scott received conflicting and confusing orders from General Charles Lee, which ultimately led to an American retreat. However, an outraged General George Washington, riding up and down the ranks, turned the troops back to the attack. General Scott later recalled that General Washington reviled Lee. "He [Washington] swore … till the leaves shook on the trees … he swore like an angel from heaven."

The British eventually withdrew during the night and retreated back to the safety of the British-held New York City.

POWHATAN TO RICHMOND

Mile Mark 80.2 — Depart Powhatan. Continue going east through town.

Mile Mark 80.5 — Reach a roundabout. Yield to traffic and go 180 degrees around the roundabout, continuing east on Route 13.

Mile Mark 82.7 — Reach the intersection with US Route 60. Turn right and rejoin Route 60 heading east.

Mile Mark 86.0 — Reach the town of Flat Rock.

Mile Mark 96.3 — Pass a series of markers that include the Salisbury and Midlothian coalmine markers.

Midlothian Coalmines

South of here are the Midlothian Coalmines, probably the oldest coalmines in America. Coal was first mined here before 1730, and during the Revolution, coal from these mines supplied the cannon foundry at Westham. The first railroad in Virginia was built from the Midlothian mines to the town of Manchester in 1831. The mines produced coal that was used in casting cannon at the Tredegar Iron Works in Richmond during the Civil War. Mining operations ceased in 1923.

Department of Conservation and Historic Resources, 1987.

(Route 60 Marker)

Salisbury

Nearby stood Salisbury, built during the middle portion of the 18th century. It was a one-and-a-half-story frame house that had two asymmetrical brick chimneys. Patrick Henry leased Salisbury from Thomas Mann Randolph and lived there while he was governor of Virginia from 1784 to 1786. Randolph sold the farm to Dr. William Turpin, and after Turpin's death his daughter Caroline, married to Dr. Edward Johnson, inherited the property. Their son, Confederate Maj. Gen. Edward Johnson, inherited the house in 1843 and lived there until his death in 1873. The house burned to the ground in 1923.

Department of Historic Resources, 2000.

(Route 60 Marker)

Mile Mark 98.6 — As traffic congestion brings progress on this road trip to a crawl, it is clear you have reached the suburbs of Richmond.

Mile Mark 101.2 — Reach an intersection with the Powhite Parkway. Continue going straight on US Route 60.

Mile Mark 105.9 — Watch for an intersection where Route 60 makes a jog. Turn left on Roanoke Street and continue following Route 60.

Mile Mark 106.3 — Route 60 turns onto Forest Hill Avenue. Turn right and continue following Route 60 to downtown Richmond.

Mile Mark 106.5 — Reach a split in the road. Bear left at the split and continue on Route 60.

Mile Mark 107.7 — Reach the intersection with US Route 1 and 301. Watch for views of the Richmond skyline up ahead in the distance.

Mile Mark 108.1 — Follow Route 60 to the left.

Mile Mark 108.3 — Re-cross the James River.

Mile Mark 108.9 — Route 60 makes a turn to the right but get into the middle lane and continue going straight onto 9th Street for a visit to downtown Richmond.

Anna Maria Lane
Soldier of the American Revolution

Near the Bell Tower in Capitol Square stood the barracks of the Public Guard. There, from 1801 to 1807, lived John Lane and his wife, Anna Maria Lane, the only documented woman veteran of the Revolutionary War to reside in Virginia. She disguised herself and enlisted with her husband in the Connecticut Continental Line. "In the garb, and with the courage of a soldier, (she) performed extraordinary military services," and was wounded at Germantown, Pa., in 1777. She followed Lane through his subsequent service in the Virginia light dragoons, and then, after the war, to the Public Guard. Granted a pension in 1808, she died on 13 June 1810.

Department of Historic Resources, 1997.

(Richmond Marker)

Mile Mark 109.1 — Watch for the Anna Maria Lane marker on the right. As noted by the marker, she disguised herself and served as a soldier in the Revolutionary War with her husband, John Lane.

Mile Mark 109.3 — Pass Richmond's Capitol Square and turn right on Broad Street. Find a parking space and walk back for a visit to Capitol Square.

Arrive in downtown Richmond.

RICHMOND
VIRGINIA

After becoming Governor of the State of Virginia, Thomas Jefferson and the Virginia General Assembly moved the state capital from Williamsburg to Richmond. They wanted the capital more centrally located and less exposed to British incursions.

However, on January 4, 1781 during the second British invasion of Virginia, General Benedict Arnold reached Richmond with about 1,000 men. The militia had, for the most part, abandoned the few defensive positions around the capital leaving it open to the British. Prior to Arnold's arrival, the Virginia government moved to the safety of Charlottesville.

From Richmond, General Arnold sent a letter to Governor Jefferson offering to spare the capital if he was allowed to seize tobacco unmolested. After Jefferson, refused, Arnold ransacked the city, destroying and burning homes and buildings as well as Government papers.

After Arnold's destructive raid, he withdrew from the city to the Virginia coast and set-up defensive positions in Portsmouth, VA.

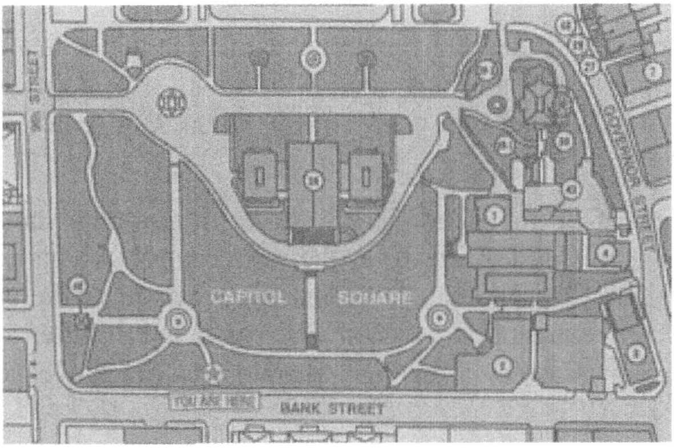

Annoyed by Arnold's raid in his home state, General George Washington sent Lafayette to Virginia to put a stop to Arnold's destruction and hopefully capture the traitor, who was now spilling American blood.

Lafayette arrived here in Richmond on April 29th, just in time to prevent a second attack on Richmond. Arnold and General Phillips had combined forces and attacked Petersburg to the south of Richmond. Petersburg was defended by Virginia militiamen under the command of Colonel J. P. E. Mullenburg. The militia held off the British for the better part of a day, but eventually retreated after being beaten by British artillery and a flanking movement.

British General Cornwallis arrived in Petersburg about a month later and took command of all British forces in Virginia. After he arrived, Arnold left his command and returned to New York City. General Phillips' forces had already lost their commander to typhoid fever. He died in Petersburg.

Cornwallis, not really interested in Richmond, headed towards Williamsburg and Yorktown. He hoped to establish a strong base in this area that could be supplied by sea.

State Capitol—Today, a visit to Richmond should begin in Capitol Square at the state capitol, which was designed by Thomas Jefferson two years after the end of the Revolutionary War. He modeled it after the Maison Carrée, a Roman temple in Nimes, France.

Inside, in the Rotunda, is a life size statue of George Washington. The statue was created by Jean Antoine Howdon, for which George Washington actually posed.

The Virginia General Assembly, which still meets here, is the oldest legislative body in America. In one of their meeting rooms is a mural depiction of the "Siege of Redoubt 10" (Redoubt 10 is an upcoming site on this road trip). The siege was led by Alexander Hamilton during the battle of Yorktown.

Virginia Executive Mansion - The mansion is home to Virginia's governor and can be found at the northeastern corner of Capitol Square. On this site at one time was the home of Patrick Henry. Henry gave the famous "Liberty or Death" speech at St. John's Church (an upcoming site on this visit to Richmond).

The current governor's mansion was completed in 1814. It is oldest, continuously occupied, governor's mansion in America.

John Marshall House — The home, built about seven years after the Revolutionary War, can be found about two blocks north of Capitol Square at the corner of 9th and Marshall Streets.

Executive Mansion

On this site lived three presidents of the United States, Monroe and Tyler as governors, and William Henry Harrison while his father, Benjamin Harrison, signer of the Declaration of Independence, was governor. The first white house on this site was also home to Governor Patrick Henry and Henry Lee, father of Robert E. Lee.

This present house, first occupied by Governor James Barbour in 1813, was designed by Alexander Parris and built by Christopher Tompkins. As the home of Virginia's chief executives since that date, it is the oldest continuously occupied official governor's residence in the nation.

In 1989, the exterior of the mansion was restored to its circa 1830 appearance under the direction of the Citizens' Advisory Council for Interpreting and Furnishing the Executive Mansion.

(Richmond Marker)

John Marshall purchased the equivalent of the entire city block in 1788. He eventually added five outbuildings to service the house. In front of you, on the present site of the John Marshall Courts Building, sat Marshall's law office, a two-story brick structure.

(Richmond Marker)

John Marshall was a Revolutionary War veteran. He began service as a lieutenant and briefly sparred with Benedict Arnold in Virginia. He fought in the battles of Brandywine, Germantown, Monmouth and Stony Point.

At Stony Point, a fort on the west bank of the Hudson River, Marshall participated in a daring assault performed at night with bayonets only — no loaded weapons. The successful attack was led in July 1779 by General "Mad" Anthony Wayne. A month later, Major Henry Lee and Captain John Marshall led a similar raid against Poules Hook, a fortified position opposite New York City on the Hudson. They successfully captured the position and took several hundred prisoners as well as military spoils.

The Residence of John Marshall

Chief Justice of the United States. Born 1755, died 1835.

Placed by the Association for the Preservation of Virginia Antiquities, 1907.

(Richmond Marker)

After the war, Marshall served the State of Virginia as an assembly member, a delegate to the convention that ratified the Constitution, and a congressman. He was Secretary of State under President John Adams and became the "Great Chief Justice" of the U.S. Supreme Court from 1801 until his death in 1835. While serving as chief justice, Marshall wrote a five-volume biography of his friend, George Washington.

Saint John's Episcopal Church — The church can be reached by car by following Broad Street east about a mile. Broad Street is the street between Capitol Square and Marshall Street. The church is at the corner of Broad and 24th Streets.

St. John's Church was built in 1741 on land donated by William Byrd II and is still an active Episcopal Church.

The church served as the site for the Second Virginia Convention during the week of March 20, 1775. The

Saint John's Episcopal Church

Here on 23 March 1775, Patrick Henry delivered his "Liberty or Death" speech, calling for American Independence, during the second Virginia revolutionary convention that included as members: George Washington, Thomas Jefferson, Peyton Randolph, and Richard Henry Lee. Saint John's Church was built in 1741 by Richard Randolph on land donated by Richmond's founder, William Byrd II. It continues to serve Henrico Parish (founded 1611). Buried in its churchyard are George Wythe and Elizabeth Arnold Poe, mother of Edgar Allan Poe.

Department of Historic Resources, 1991.

(Richmond Marker)

convention was held here because the current royal governor of Virginia, Lord Dunmore, was unable to maintain control over the Virginia Assembly in Williamsburg and suspended it. Among the delegates at the convention here at St. John's Church was the great orator, Patrick Henry.

Patrick Henry embodied the spirit of American courage and patriotism. He is recognized today, as he was among his contemporaries, as the orator of liberty. His compelling speeches kindled the fires of the Revolution and fueled the effort to secure freedom.

Henry's was the first voice raised against England in its attempt to impose taxation without representation. He rose to his full stature in attacking the infamous Stamp Act, which was hotly debated at the House of Burgesses in Williamsburg in 1765. The other delegates shouted in support when Henry hurled defiance at George III with the challenge, "If this be treason, make the most of it."

Patrick Henry's most famous speech was delivered during the Second Virginia Convention. His words became the clarion call that led the colonies into the Revolution. With courage and eloquence, he declared:

> "Gentlemen may cry peace, peace, but there is no peace. The war is actually begun! The next gale that sweeps from the north will

"Give Me Liberty or Give Me Death"

To the glory of God and as a grateful tribute to the memory of her illustrious son and first governor, Patrick Henry, patriot-orator-statesman, the Commonwealth of Virginia has caused this tablet to be erected.

In the Virginia Convention assembled in this church, March 28, 1776, by his immortal eloquence he inspired in his countrymen the clear vision of truth and duty and roused them to consecrate themselves to the defense of liberty

(Richmond Marker)

bring to our ears the clash of resounding arms. Our brethren are already in the field! Why stand we here idle?"

"What is it that gentlemen wish? What would they have? Is life so dear or peace so sweet as to be purchased at the price of chains and slavery? Forbid it, Almighty God -- I know not what course others may take; but as for me -- give me liberty, or give me death!"

Henry's leadership did not end once the Revolution was won. Perhaps his greatest contribution to the nation was in working toward the adoption of the Bill of Rights. While desiring a more effective federal

This tablet is dedicated to mark the site where lie the mortal remains of George Wythe. Born 1726. Died 1806. Jurist and statesman. Teacher of Randolph, Jefferson and Marshall. First professor of law in the United States. First Virginia signer of the Declaration of Independence.

Erected by Patriotic Citizens of Virginia, A.D. 1922.

(Richmond Marker)

government, he was adamant in demanding protection of basic individual civil liberties.

The first American governor of Virginia, Henry served five exhausting terms. In 1794, he retired to his last home and burial place, Red Hill, near Lynchburg, Virginia.

Today, St. John's is open for guided tours seven days a week, except on Easter

Sunday, Thanksgiving Day, Christmas Eve, Christmas, New Year's Eve and New Years Day. The tours are held from 10 AM until 3:30 PM Monday through Saturday and 1 PM until 3:30 PM on Sunday. A donation is requested of all taking the guided tour.

Reenactments of the Second Virginia Convention are held every Sunday afternoon promptly at 2 PM, beginning the last Sunday in May and continuing through the first Sunday in September. In addition, a commemorative reenactment is held on the Sunday that falls closest to March 23, the anniversary of Patrick Henry's famous speech.

RICHMOND TO ST. PETER'S CHURCH

Mile Mark 110.3 — Depart St. John's Church heading east on Broad Street, which is also US Route 60 East.

Mile Mark 110.7 — Pass the Chimborazo Visitor Center, which is part of the Richmond National Battlefield Park. About 80 years after the Revolutionary War, Richmond was under attack again, this time by the Union Army during the Civil War. Richmond was the capital of the Confederacy and home to its President, Jefferson Davis.

This area includes part of the outer defenses for Richmond during the Civil War.

Mile Mark 110.9 — Route 60 bears to the right and jogs to the left as you clear the battlefield area.

Mile Mark 112.1 — Reach the intersection with Williamsburg Road as you approach a Civil War burying ground. Turn left and continue east on Route 60.

Mile Mark 115.4 — Pass the Richmond Airport on the right.

Mile Mark 116.5 — Watch for the McClellan's Picket Line Markers. This area was once held by the Union Army during the Civil War.

Mile Mark 118.8 — Reach the intersection with Interstate 295, Richmond's outer beltway.

Mile Mark 122.5 — Watch for a series of markers on the right, one of which is the New Kent County marker. This marker summarizes the history of the county and upholds a tradition that George Washington and Martha Custis were wed at St. Peter's Church, the upcoming site.

The marker also mentions the county's association with Virginia's early involvement in the Revolutionary War. In early 1775, the Royal Governor of Virginia, Lord Dunmore, was alarmed by the military preparation that was going on in his province. He cautioned the earl of Dartmouth that this preparation might be "employed

New Kent County

Established in 1654 from York and named for County Kent, England. In it are Chestnut Grove, Birthplace of Martha (Dandridge) Custis; White House, home of Martha Custis; Poplar Grove, where Martha Custis met George Washington; St. Peter's Church, where Washington was married to her; Eltham, where John Parke Custis, Washington's stepson, died in 1781; Barhamsville, where Patrick Henry received money to pay for powder seized by Lord Dunmore, 1775; Providence Forge, remains of one of the oldest iron forges in America. New Kent County was traversed by Cornwallis and Lafayette in 1781.

Erected by the New Kent County Citizens' Association 1931.

(US Route 60 Marker)

against the government, if occasion require." On April 21st, he seized the Colony's cache of gunpowder from the magazine in Williamsburg. His action, although in his own defense, greatly angered the Colonists, and an armed force of several thousand, led by Patrick Henry, marched on Williamsburg to force the governor to return the cache. Dunmore refused to do this, but he did agree to pay for the gunpowder in full. As noted by the New Kent County marker, it was near this area where Patrick Henry received this compensation. However, Dunmore, still fearing the anger of the Colonists, fled Williamsburg and took refuge on a British man-of-war in the harbor. Not to be defeated, Dunmore would continue to harass the Colonists from the sea over the next several years.

The New Kent Road marker records the passing of Cornwallis' army six years later. He marched his army numbering about 7,000 to Williamsburg passing through this area.

These markers not only indicate this area was a crossroad during the Revolutionary War, but also a crossroad 80 years later during the Civil War.

Mile Mark 123.4 — Reach the intersection with Route 249. Turn left on Route 249 heading east.

New Kent Road

This was the main road to Williamsburg in early days. Cornwallis, retiring eastward, used this road in June 1781. The confederates, retreating westward, passed over it in May 1862.

Conservation & Development Commission 1937

(US Route 60 Marker)

Mile Mark 123.7 — Go over Interstate 64.

Mile Mark 130.2 — Watch for the White House and St. Peter's Church markers on the right.

Mile Mark 130.3 — Reach the intersection with Old Church Road. Turn left heading north on Old Church Road.

Mile Mark 131.2 — Reach the intersection with St. Peters Lane. Turn right heading east on St. Peters Lane.

The White House

This place, six miles northeast, was the home of Martha Custis. According to tradition, George Washington first met her at Poplar Grove, nearby, in 1758. On January 6, 1759, Washington and Martha Custis were married, it is believed at the White House. The estate descended to W. H. F. Lee, son of Robert E. Lee. The house was burned by Union Troops when McClellan made the White House his base of operations in May 1862.

Conservation & Development Commission 1930.

(Route 249 Marker)

St. Peter's Church

Two miles northeast is St. Peter's Church, built in 1703 in English bond. David Mossom, rector there for forty years, was the minister who married George Washington. According to one tradition, the wedding took place at St. Peter's Church.

Conservation & Development Commission, 1930.

(Route 249 Marker)

Mile Mark 131.8 — Bear left and go through the gate of St. Peter's Church.

Mile Mark 133.9 — Arrive at St. Peter's Church.

ST. PETER'S CHURCH
NEW KENT, VIRGINIA

St. Peter's Parish Church is among the oldest congregations in America with its establishment in 1679. The current church building was begun in 1701 and was in use by 1703. St. Peter's is the oldest church in the state.

The church is often called "the first church of the first lady" because Martha Dandridge, who would later become the wife of George Washington, was a member. In June 1749, she had originally married Colonel Daniel Parke Custis, a former churchwarden and vestryman, but was widowed by 1757.

Two years later, Martha would marry Colonel George Washington on January 6, 1759. The church rector, the Reverend David Mossom, presided over the ceremony.

St. Peter's Parish Church (Episcopal)

Parish created, 1679.
Church erected, 1704.

The first church of the first First-Lady.

(St. Peter's Marker)

The following is a translation of the Latin inscription on the tombstone of the Reverend David Mossom. This tombstone, now partially destroyed, is located in the northeast corner of the inside wall of this church. The inscription indicates that the Reverend David Mossom is buried beneath the church chancel … The Reverend David Mossom baptized Martha Washington (Dandridge). He officiated at her marriage to Daniel Parke Custis in June 1749; later, he officiated at her marriage to George Washington, January 6, 1759.

(St. Peters Marker)

Today, St. Peter's remains an active church with two services held each Sunday (9 & 11 AM). In 2003, it celebrated the tercentennial of the construction of the current building.

ST. PETER'S CHURCH TO COLONIAL WILLIAMSBURG

Mile Mark 133.9 — Reverse direction and depart St. Peters Church.

Mile Mark 132.6 — Reach the intersection with Old Church Road. Turn left heading south.

Mile Mark 133.5 — Reach the intersection with Route 249. Continue across the intersection heading south.

Mile Mark 135.0 — Go over Interstate 64.

Mile Mark 138.1 — Reach the intersection with US Route 60. Just to the west of this intersection on the south side of Route 60 is the Long Bridge Marker.

The Long Bridge marker records the landing of the second invasion of Virginia about 20 miles south at Westover Plantation on the James River in January of 1781. It was from this plantation that General Benedict Arnold marched on Richmond, but as noted by the Long Bridge marker, he sent Colonel Simcoe and his Queen Rangers north of the plantation. Their mission was to destroy the Colonists war-making industry, including a gunpowder factory and an iron foundry. At

Long Bridge

One mile south is Long Bridge over the Chickahominy River. Benedict Arnold sent Simcoe there in the British invasion of 1781. Longstreet crossed there in the Peninsular Campaign, May 1862. Grants Fifth and Second Corps crossed there, in June 1864, on the way to Petersburg.

Virginia Conservation Commission 1947.

(US Route 60 Marker)

Charles City, they cornered a Virginia militia contingent, killing about 20 Americans.

Mile Mark 143.5 — Reach the intersection with Route 157. Watch for the Providence Forge marker on the right.

As noted by the marker, Lafayette moved his forces through this area shadowing Cornwallis. Cornwallis

Providence Forge

Here about 1770, Charles Jeffery Smith, a Presbyterian Minister, settled and, with William Holt, built a forge for making farm implements. Francis Jerdone became a partner in 1771. A militia camp was established here in 1781, and Lafayette was here in July and August 1781.

(Route 60 Marker)

was on the move to Portsmouth, about 20 miles to the east, carrying out orders from General Clinton to return 3,000 of his troops for the defense of New York City. Lafayette's forces now numbered about 5,000 thanks to reinforcements sent by George Washington.

Mile Mark 156.7 — Reach the intersection with Route 30. Turn right and continue following Route 60 east.

Mile Mark 157.2 — This part of Route 60 is also called the Washington/Rochambeau Route as noted by the marker on the right.

In a brilliant military decision, Washington quickly brought an Allied army, consisting of American and French forces from their offensive positions around New York City to this area in late September 1781 to attack Cornwallis. The French forces were under the command of General Comte de Rochambeau.

In early August, prior to the arrival of the Allied army, Cornwallis had chosen Yorktown as his headquarters when he was directed to occupy a deep-water port that could be accessed by the British fleet. His orders to send 3,000 troops to New York from Portsmouth were countermanded.

While the allied army was on the move, a large French fleet under Admiral Comte de Grasse had sailed up from the West Indies and proceeded to blockade the mouth of the Chesapeake Bay. In the Battle of the Capes, de Grasse won a victory over Admiral Thomas Graves and his British fleet and on September 5th, proceeded to maintain a strict blockade, cutting off Cornwallis from help or escape by sea.

Mile Mark 157.3 — Watch for the Hickory Neck Church and marker on the left. The marker indicates that on April 21, 1781 Patriot militia camped on the church grounds. The militia may have been in the area

Hickory Neck Church

Hickory Neck Church was built about 1740. Militia opposing the British camped here on April 21, 1781. A few miles north is the foundation of an ancient stone house, dating possibly from about 1650.

Virginia Conservation Commission 1848

(Route 60 Marker)

in response to British General Philip's third invasion in March 1781.

Mile Mark 158.4 — On the right is a Chickahominy Church marker indicating that Lafayette's forces camped on the church grounds on July 6-8, 1781.

When Cornwallis was ordered to Portsmouth, he had to cross the James River and he was well aware of the dangers of river crossings, especially with Lafayette in the area. In an attempt to draw the American army into an ambush, Cornwallis reversed the usual river-crossing procedures and sent his army's baggage first keeping the bulk of his forces hidden on the north side of the river.

Lafayette did move in to attack what he thought was the British rear guard. His advance American guard of about 500 men under the command of Anthony Wayne began to skirmish with the British forces while Lafayette moved his troops into the area. However, Lafayette didn't fall into the trap. He personally

Chickahominy Church

Two miles south is the site of the Colonial Chickahominy Church now destroyed. Lafayette's forces camped there, July 6-8, 1781. The church was used as a hospital after the Battle of Green Spring, July 6, 1781.

Conservation & Development Commission 1927

(Route 60 Marker)

reconnoitered the British placement of troops and discovered the plot, but it was too late to warn Wayne.

Cornwallis attacked Wayne at Green Spring with the full weight of his British forces. Caught in the middle of an overwhelming ambush, Wayne led his tiny American army on a bayonet counter-attack into a line of British troops facing their musket fire, grapeshot and overwhelming odds. His furious counter-attack stopped the British assault and he was able to escape with the bulk of his forces.

In the battle, the British had about 75 casualties and the Americans twice that. Amongst the American casualties, 28 were killed.

Mile Mark 162.5 — Watch for a series of markers on the right, one of which is the Green Spring marker. Five miles south of this marker is the area where the Battle of Green Springs took place. Today, the area and

Green Spring

On this road, five miles south, is Green Spring, home of Governor Sir William Berkeley. Bacon the Rebel occupied it in 1676. Cornwallis, after moving from Williamsburg by this road on July 4, 1781, was attacked by Lafayette near Green Spring on July 6, 1781. Anthony Wayne was the hero of this fight.

Conversation Development Commission 1927

(Route 60 Marker)

Six-Mile Ordinary

Six-Mile Ordinary, a popular 18th-century tavern also known as Allen's for its proprietor Isham Allen, stood six miles from Williamsburg. On 1 July 1774, a group of free holders congregated there and drafted the James City Resolves not to import British goods. Two years later, they gathered again to declare their support for American independence. On 21 April 1781, Colonel James Innes notified the governor that 500 British infantrymen, 50 horses, and 4 pieces of artillery had come ashore at Burwell's Ferry. Because of this unexpected event, Innes and his troops retreated to Six-Mile Ordinary around midnight.

Department of Historic Resources, 1998.

(Route 60 Marker)

its architectural remains are protected by the National Park Service.

Also among the series of markers is the Six-Mile Ordinary marker and the Spencer's Ordinary marker.

Spencer's Ordinary

On this road, four miles south, the action of Spencer's Ordinary was fought, June 24, 1781, between detachments from Lafayette's and Cornwallis' armies.

Virginia Conservation Commission 1950.

(US Route 60 Marker)

Mile Mark 165.1 — Reach increased congestion and slow travel as you pass major shopping centers near Colonial Williamsburg.

Mile Mark 166.5 — Route 60 bears to the left. This is also the route to the Colonial Williamsburg Visitor Center. For a short visit, it is recommended that you keep to the right so you can find street parking for a stroll around Colonial Williamsburg. However, to suggest that one could make a short visit to this area, which is so rich in history, might be considered blasphemy. The reader is reminded of the caveat printed on the back cover: "With so much to see, it could take a week."

Mile Mark 166.8 — Get into the right lane as you approach Lafayette Street and Monticello Avenue.

Mile Mark 167.6 — Pass the Admissions Building of the College of William & Mary. Find a parking spot among the two-hour parking spots that can be found on both sides of the street.

Mile Mark 168.2 — Arrive in Colonial Williamsburg.

COLONIAL WILLIAMSBURG
WILLIAMSBURG, VIRGINIA

Williamsburg was the capitol of Virginia for over 80 years from 1699 to 1780. Prior to that time, the capital was in Jamestown. After the Jamestown statehouse burned for the fourth time, the Royal Governor at the time, Francis Nicholson, seized on the opportunity to propose that the capital be moved to a better location further inland.

The House of Burgesses chose the town called Middle Plantation, which they renamed Williamsburg, in honor of the king of England, William III. At the time, the town contained a church, a college, a tavern, two mills, several stores and neighborhoods of homes, many belonging to local tobacco planters.

Today, Colonial Williamsburg is one of the oldest and largest living history museums. It is the restored and reconstructed capital of eighteenth-century Virginia.

The College of William and Mary — Begin a walking tour of Williamsburg at the eastern end of the college. In this area is the Wren Building, the oldest academic building in use in America. Flanking this

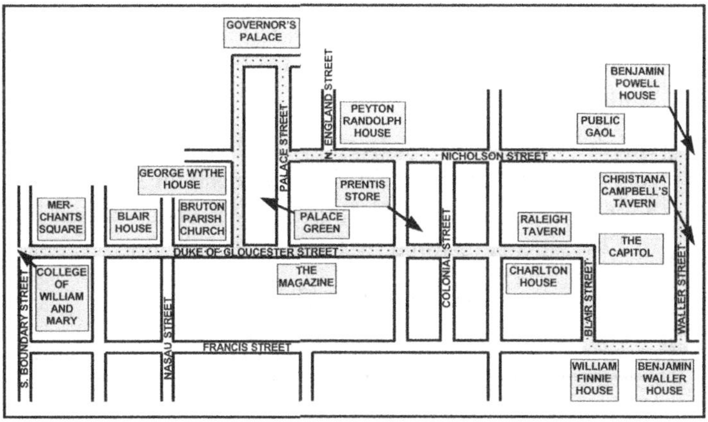

The College of William and Mary in Virginia

A charter was granted by King William and Queen Mary in 1693. By fostering "good arts and sciences" and by educating the youth in "good letters and manners," the college has maintained its original mission as "a place of universal study."

The college was transferred to the Commonwealth of Virginia in 1906.

The main structure, now called the Wren Building erected in 1695, burned in 1705, was rebuilt in 1716 on the original foundation. After several fires and alterations, it was restored in 1928 by Mr. John D. Rockefeller, Jr. The General Assembly met in this building 1700 to 1704, and again 1747 to 1754. It is the oldest academic building in continuous use in the United States.

On the left is the Brafferton, erected as an Indian school in 1723. On the right is the President's House, built in 1732, accidentally burned by French soldiers in 1781, it was restored by Louis XVI in 1786.

Gift of Class of 1956

(Williamsburg Marker)

building to the north is the President's House and to the south is the Brafferton.

Construction of the Wren Building began in 1695, two years after the college received its charter from King William and Queen Mary. Several of the college's distinguished alumni have attended classes in the Wren Building, including Thomas Jefferson, John Marshall and James Monroe.

A chapel was added to the Wren Building in 1732. In the crypt below the chapel are the remains of Sir John Randolph and his sons Peyton and John. Peyton and John, who were cousins of Thomas Jefferson, split loyalties. John, despite his close kinship with Jefferson, regarded the call for independence as treasonous. Peyton, on the other hand, was so supportive of the American cause, he was eventually elected speaker of the Continental Congress.

The President's House, as the name infers, is the residence of the college president. However, it was built in 1733 and was never residence to the College's first president, Reverend James Blaire, its founder. During the Revolutionary War, the house served as headquarters to General Cornwallis and General Rochambeau.

The Brafferton was built in 1723 with support from the manor of Brafferton in Yorkshire, England. Until the Revolutionary War severed that support, the

Thank You

By purchasing an admission ticket, you have helped preserve the historic area for future generations.

Colonial Williamsburg is a private, not-for-profit educational organization. Income from admissions and contributions provides the funds needed to maintain and preserve the historic area.

(Williamsburg Marker)

building was used for the education and conversion of Indian boys. Prior to the war, a dozen or so Indian boys could be found learning and playing around campus.

John Blair House and Kitchen — The home can be reached by walking east on Duke of Gloucester Street, which is opposite the entrance to the College of William and Mary and the Wren Building. As you pass Merchants Square and enter the historic area, there is a small building on the south side of Duke of Gloucester Street where visitors may purchase an admission ticket. A ticket is not necessary for a short walking tour, but it is recommended. Your ticket not only is required to see many of the attractions, but it is also a visible show-of-support for the programs and activities of Colonial Williamsburg.

Once inside the historic area, the Blair House is about a half-block on the north side of Duke of Gloucester Street. John Blair, Sr., whose uncle was the Reverend James Blair, was a burgess and auditor for the colony of Virginia from 1728-1771. His son, John Blair, Jr., graduated from the College of William and Mary and served as a burgess during the Revolutionary War. After the war, he served as a Virginia representative at the Constitutional Convention and was appointed to the Supreme Court by President George Washington.

Bruton Parish Church

Services on Sunday & Special Days.

One of the oldest Episcopal Churches in America, this historic building has been in service to God and Man continuously since 1715.

Open to visitors every day.

(Williamsburg Marker)

Bruton Parish Church — The church can be found about a block past the John Blair House on Duke of Gloucester Street. The congregation of the church was founded from several smaller congregations in 1674. The current church was completed in 1715 and has since been in continuous use.

Among the parishioners attending the church were George Washington, Thomas Jefferson and James Monroe. Also among the parishioners were black slaves who were welcomed in the church. George Washington personally sponsored more than a dozen slaves for baptism, serving as Godfather.

During the Revolutionary War, the parish church bell served as the town's liberty bell.

George Wythe House — The home can be found by following Duke of Gloucester Street to the intersection with Palace Street on the west side of the Palace Green. The home is a block north of Duke of Gloucester Street and requires an admission ticket for entry.

George Wythe was among the most influential Virginians during the Revolutionary War. After Thomas Jefferson attended William and Mary, he studied law under George Wythe. Jefferson called him "my faithful and beloved mentor in youth, and most affectionate friend through life."

Wythe served in the House of Burgesses and was a delegate to the Continental Congress. He was also a signer of the Declaration of Independence. In 1779, he became the first Chair of Law at William and Mary and among his students were James Monroe and John Marshall.

In 1781, the Wythe House served as headquarters for George Washington. After the surrender at Yorktown, the house was headquarters for General Rochambeau.

George Wythe died in 1806 and is buried at St. John's Church in Richmond (a previous site on this road trip).

The Governor's Palace — The palace is the great structure at the north end of the Palace Green. Completed in 1722, it was at its time, one of the finest structures in America. Today, the reconstructed palace lives up to its 1722 reputation. An admission ticket is required for a tour.

For most of Williamsburg's colonial history, the palace was home to the Royal Governor, but beginning in 1765 a movement began in Virginia that would eventually move the palace out from under the British Crown and place it in the hands of the colonists. By April of 1775, the Royal Governor at that time, Lord Dunmore, clearly feared for his and his family's safety. Despite efforts by Virginia's moderate leaders to keep

the governor in his palace, Dunmore and his family fled Williamsburg by sea on June 8, 1775. He later retaliated by ordering the shelling of Norfolk on January 1, 1776, burning the town in a great fire.

Beginning in July 1776, the palace served as residence for the first two Colonial governors of Virginia, Patrick Henry and Thomas Jefferson. After Jefferson moved the capital to Richmond and during Cornwallis' brief stay in Williamsburg, the palace served as a hospital. Behind the palace is a graveyard where over 150 American soldiers are buried.

On the night of December 22, 1781, fire broke out in the palace and it burned to the ground. Today's reconstruction is built on the foundations of the original palace.

Visitors to the palace are met at the door by a guide in period dress and are shown throughout the grand structure. British royalty and the power of the Crown are emphasized in the tour as over 800 guns and swords decorate the walls and ceilings of many of the rooms. Among the furnishings are several pieces actually owned by Lord Dunmore.

Peyton Randolph House — From the palace, the Randolph house can be reached by following the eastern side of Palace Street south past the Play Booth Theatre and turning east on Nicholson Street. The Peyton Randolph House is on the left at the corner of North England and Nicholson Streets.

Peyton Randolph, as mentioned earlier, was the son of Sir John Randolph, and was a patriot for the American cause. Besides serving as speaker of the Continental Congress, he also served Virginia as attorney general and speaker of the House of Burgesses. He died of a stroke at the beginning of the Revolution in Philadelphia in 1775.

Peyton Randolph House

For more than fifty years this was the home of Peyton Randolph (1721-1775), who served the Colony of Virginia in many of its highest governmental offices and became the first president of the Continental Congress. His father, Sir John Randolph, the only colonial Virginian to be knighted, lived here until his death in 1737. The original house is unusual for its seven fully paneled rooms. The furnishings reflect the home of a family prominent in the social and political life of the colony.

(Williamsburg Marker)

After his death, his wife, Mrs. Betty Harrison Randolph, continued to live in the house. However, prior to the Siege of Yorktown, she relinquished the house to General Rochambeau, who also made this house his headquarters for a period of time.

Public Gaol — The gaol (jail) can be found about five blocks east of the Peyton Randolph House on the north side of Nicholson Street.

The gaol, which is behind the Capitol building, served as a military prison during the Revolutionary War. Confined to the gaol were captured British soldiers, Tory sympathizers, traitors, deserters and spies.

The Public Gaol

This was Virginia's chief prison which housed debtors and criminals and served as the jail for the General Court in the nearby capital. Here Blackbeard's pirates, captured in 1718, were confined until the day of their hanging. Leg irons, an exercise yard, food slots, and criminal cells with primitive sanitation have been restored to their early appearance.

(Williamsburg Marker)

After the war, the building continued to serve as a Williamsburg prison into the twentieth century. Today, the restored gaol is furnished with shackles unearthed during its restoration.

Benjamin Powell House — The house can be found opposite the end of Nicholson Street on the east side of Waller Street.

Powell was a builder and among the structures he erected was the tower and steeple at Bruton Parish Church. Powell was a patriot and served on an embargo committee of British goods. Other members included George Wythe and Peyton Randolph.

Christiana Campbell's Tavern — The tavern can be found about a block south of the Powell House. The tavern was one of George Washington's favorite dining spots. Today, the reconstructed tavern continues its colonial tradition offering many colonial dishes on the menu. The tavern is a recommended spot for dinner.

Benjamin Waller House — The house can be found at the south end of Waller Street, just across the intersection of York and Francis Streets.

Waller was a prominent Williamsburg attorney. George Wythe was among the law students that he mentored.

About a half-block west of the Waller House is an entrance that leads to Bassett Hall, the former Williamsburg house of John D. Rockefeller. Rockefeller is largely responsible for providing capital to support the reconstruction and restoration of Colonial Williamsburg, which began in 1926.

William Finnie House — The house can be found about a block west of the Benjamin Waller house. The architecture of the house is considered classic Williamsburg.

During the Revolutionary War, William Finnie served the southern department of the Continental army as Quartermaster.

The Capitol — The Capitol is located at the corner of Blair Street and the east end of Duke of Gloucester Street. Blair Street is about a half-block west of the William Finnie House.

The Capitol is a reconstruction of Virginia's first capitol in Williamsburg. Fire gutted the first capitol in 1747 and was replaced with a different structure in 1753. This second structure was destroyed by fire in 1832.

Within the second capitol, many major events took place that led Virginia and the other colonies to war with the British. The first event took place on May 30, 1765 during a stormy session of the House of Burgesses. A political newcomer, Patrick Henry, introduced several resolutions attacking the Stamp Act (a tax on various commercial and legal documents) and the notion of taxation without representation. His resolutions, copies of which spread throughout the colonies, influenced mob violence in Boston and the creation of the Stamp Act Congress in New York City.

British Parliament, under pressure from commercial interests, repealed the Stamp Act in February 1766. However, in June of 1767, King George III assented to a new series of taxes on imports including glass, paint, paper and tea (Townshend Acts). Four years later, again under pressure from commercial interests, the Parliament repealed all import duties on all items except tea.

The Virginia Assembly, meeting in the Capitol, with members now including Thomas Jefferson, George Washington and Richard Henry Lee, passed a resolution that June 1, 1774 would be set aside as a day of fasting, humiliation and prayer in support for its "sister colony," Massachusetts, where protests such as the famous Boston Tea Party were escalating to a violent confrontation with British forces.

Two days after the resolution was passed, Virginia's Royal Governor, Lord Dunmore dissolved the Virginia Assembly. Despite the governor's action, the assembly reconvened at the Raleigh Tavern, just west of the Capitol on Duke of Gloucester Street. Here, they pronounced that an "attack on one of the colonies is an attack on all."

In August, Virginia's leadership elected delegates to the first Continental Congress. One of those delegates, George Washington, declared, "If need be, I will raise one thousand men … and march myself at their head for the relief of Boston."

On May 15, 1776, the legislature instructed its delegates in Philadelphia "to declare the United Colonies free and independent states, absolved from all allegiance to, or dependence upon, the Crown or Great Britain …," leading directly to the Declaration of Independence. Also, during the same time period, the assembly adopted a Declaration of Rights, which later became the basis for the Bill of Rights in the US Constitution.

On July 6, 1776, Patrick Henry was inaugurated as the first elected American governor of the state of Virginia. Meanwhile, not to take his unseating lightly, Lord Dunmore, now living on British ships on the Chesapeake Bay, led a raid up the Potomac River threatening George Washington's Mount Vernon home. However, Lord Dunmore abandoned his efforts in August 1776.

On January 2, 1778, with the Revolutionary War mostly being fought in the north, a secret mission was ordered by Governor Patrick Henry in the capitol building. The great Indian fighter, George Rogers Clark was authorized to raise several companies with the mission to defend the Kentucky frontier. Under a secret mission, he also planned to attack settlements and forts along the Mississippi with his final objective being British-held Fort Detroit. At the time,

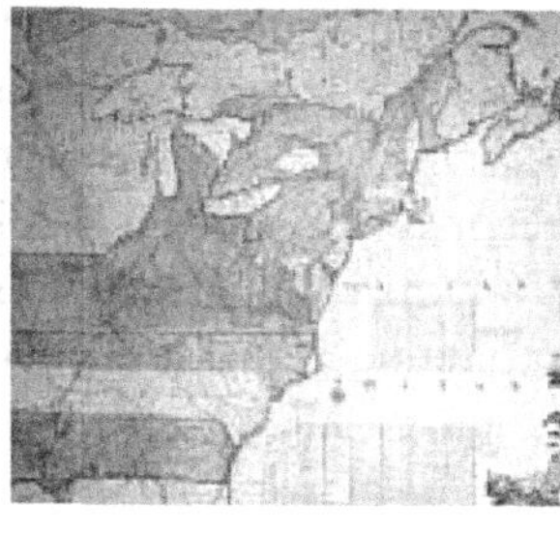

Virginia claimed land stretching west to the Mississippi and north to the Great Lakes.

When the British began to turn their attention from the north to the south, they staged their first invasion of Virginia in 1779 attacking the nearby towns of Portsmouth, Norfolk and Suffolk. The proximity of the attack clearly showed the vulnerability of Williamsburg and the legislature moved the capital from Williamsburg to Richmond the next year.

Today, visitors to the reconstructed, first Williamsburg capitol building can tour the House of Burgesses, the General Court, Council Chambers and Committee rooms.

Raleigh Tavern — The reconstructed Raleigh Tavern is about a block west of the Capitol and is on the north side of Duke of Gloucester Street.

The tavern, with its close proximity to the Capitol, served as a center for social, commercial and political gatherings. In 1775, there was a welcome home reception at the tavern when Peyton Randolph returned as president of the first Continental Congress. The tavern was also one of George Washington's favorite dining spots.

The tavern also served as the default home of the Virginia Assembly after it was dissolved by the Royal Governor. The first meeting at the tavern was in 1769 when then Governor Botetourt dissolved the assembly for protests against the Townshend Acts. The assembly reconvened at the tavern to draw up a boycott of British goods. Five years later, the assembly reconvened here when Lord Dunmore dissolved the governing body.

Today, visitors can tour the reconstructed tavern room and purchase baked goods from the tavern bakery.

Charlton House — Across the street from the Raleigh Tavern is the Charlton House. Edward Charlton, a wigmaker, performed a very important service for colonial politicians and aristocrats. Among his customers were Thomas Jefferson, Patrick Henry, George Wythe and Peyton Randolph.

Prentis Store — The store can be found about two blocks west of the Charlton House at the intersection of Colonial Street and Duke of Gloucester Street. The store is in the northwest corner of the intersection.

The store was built during the late 1730's. In 1774 Yorktown patriots staged their own "tea party." The tea they threw into the York River was consigned to the Prentis Store.

The Magazine — The magazine is about two blocks west of the Prentis Store. It is on the south side of Duke of Gloucester Street.

The magazine, which stored several tons of gunpowder, was protected by a high wall. It was on April 21, 1775 that Lord Dunmore ordered the removal of the powder from the magazine. The removal was accomplished at night.

Palace Green — The Palace Green completes the walking tour of Williamsburg. It can be found a block west of the Magazine on the north side of Duke of Gloucester Street.

The Palace Green was witness to the arrival of both British and American forces prior to the Battle of Yorktown. Cornwallis and his combined armies arrived in Williamsburg on June 25, 1781. He occupied the city for about ten days before his departure to Portsmouth and his skirmish with Lafayette's forces.

After Cornwallis moved to Yorktown, Washington and Rochambeau arrived at Williamsburg in mid-September 1781. Amid the pomp and circumstance of a grand military review, they planned the siege of Yorktown. On September 28th, the forces moved out of the city to take their positions around Yorktown.

The stage was set for the final major battle of the Revolutionary War in America.

WILLIAMSBURG TO THE YORKTOWN BATTLEFIELD

Mile Mark 168.2 — Depart Williamsburg on South Boundary Street, which separates the College of William and Mary and Merchants Square.

Mile Mark 168.3 — Reach the intersection with Francis Street. Turn left on Francis Street.

If you continue straight on Francis Street, you will reach Route 60 East. About six miles past Williamsburg, just off Route 60, is Carter's Grove Plantation. During the early 1750's, Carter Burwell erected a Georgian mansion on a 1,400-acre tract on the James River. Today, Carter's Grove is a museum exhibiting four centuries of life in Virginia. The site includes an archeological museum, the Wolstenholme Towne site, the Carter mansion, the McCrea stable and slave quarters. This site is included with the price of an admission ticket to Williamsburg.

About 20 miles further east on Route 60 are the cities of Norfolk and Portsmouth. Both cities were subjected to the first three British invasions of Virginia.

The Bombardment of Norfolk, 1776

The bombardment and burning of Norfolk helped fuel the American Revolution. In retaliation for Norfolk's refusal to supply provisions for his ships, the British Royal Governor, Lord Dunmore, ordered his fleet to fire on the town in the afternoon of New Year's Day 1776. The entire waterfront of the borough was under siege from the man-of-war HMS Liverpool lying in the Elizabeth River near hear and from six other vessels, including the HMS Otter and the HMS Kingfisher, extending westward to Town Point. One of Dunmore's cannonballs remains lodged in the wall of St. Paul's Church three blocks north. Landing parties were sent ashore to burn buildings used by snipers shooting at the ships. Dunmore's troops destroyed 19 buildings. Partly to prevent the British from occupying the borough's strategic location, Virginia and Carolina militiamen went on a rampage and further burned and plundered Norfolk over the next two days. When Colonel William Woodford and General Robert Howe had regained control of their patriot troops, about two-thirds of Norfolk, 863 buildings, had been destroyed.

View of part of the ruins of Norfolk, by Benjamin Latrobe, 1796: Courtesy of the Maryland Historical Society.

(Norfolk Marker)

Revolutionary War at Portsmouth

In October 1775, Virginia's last Royal governor, the Earl of Dunmore, made his headquarters at Cosport, one mile south of here. After his defeat at Great Bridge and the destruction of Norfolk, he entrenched at Hospital Point, one mile north, but was driven out in May 1776. Portsmouth was again invaded by the British: Sir George Collier in 1779, General Leslie in 1780, the traitor Benedict Arnold and General Phillips in 1781. Here, on this Portsmouth waterfront in August 1781, Lord Cornwallis embarked his 7,000 troops and sailed to Yorktown where he surrendered to the victorious American and French forces, 19 October 1781.

(Portsmouth Marker)

Mile Mark 168.5 — Watch for Nasau Street, which is just past the old hospital on the right. Turn right on Nasau Street.

Mile Mark 168.7 — Make the next left onto Newport Avenue.

Mile Mark 168.8 — Go over the bridge that spans the Colonial Parkway and turn left, following the signs to the Williamsburg Visitor Center and Yorktown.

The opposite direction on the Colonial Parkway goes to Jamestown on the James River. The town itself no longer exists except for what is now maintained as an archeological museum. The exhibits present Jamestown's short history and its struggle to overcome hunger, sickness and the ever-present wilderness. The centerpiece of all the exhibits is the Old Church Tower, the only 17th century structure that remains today.

Mile Mark 169.0 — Go through a tunnel that goes under Colonial Williamsburg.

Mile Mark 169.5 — Pass the entrance to the Colonial Williamsburg Visitor Center and continue going north toward Yorktown.

Mile Mark 171.5 — Go under Interstate 64.

Mile Mark 172.8 — As you approach the York River, there are several areas where you can pull over and learn about the history of the area. Most of the markers predate the Revolutionary War.

Mile Mark 175.0 — Cross Kings Creek. Watch for views of the York River on the left.

Cheatham Annex

The piers and structures across the water are an extension of the Naval Base at Norfolk. This takes advantage of the excellent York River deep-water channel as did Cornwallis when, in 1781, he chose Yorktown as his base.

(Colonial Parkway Marker)

Mile Mark 176.5 — On the left are several pullovers and views of the Cheatham Annex, an extension of the Naval Base at Norfolk.

Mile Mark 180.3 — Watch for the Route 17 Bridge across the York River and the entrance to the Yorktown Victory Center on the left. The center has several exhibits, many of which include original artifacts from the Colonial period. It also has two living-history areas: a small encampment with Continental army re-enactors in period dress and a 1780's farm. Both areas include hands-on activities.

Mile Mark 181.3 — Go over US Route 17.

Mile Mark 181.7 — Reach a stop sign at the entrance to the Yorktown National Historic Park. Continue across the busy intersection and enter the park.

Mile Mark 182.0 — Reach the visitor center of the Yorktown National Historic Park.

Converging on Yorktown

"This place is in no state of defense. If you cannot relieve me very soon, you must be prepared to hear the worst." — Lord Cornwallis, British commander at Yorktown, September 17, 1781.

"The present moment … will decide American independence and glory and superiority of the Allies." — George Washington, General Orders near Yorktown, September 20, 1781.

In 1781, France became an ally of the United States and the Revolution soon grew into an international conflict that eventually involved most of the major nations of Western Europe. The wider scope of the war forced Great Britain to fight on several fronts, as well as to contend with French naval power. Nonetheless, by 1781 the prospects for American independence still looked bleak. The British had overrun part of the south and still held New York, while Washington's army in the north had been shaken by two unsuccessful mutinies. The tide, however, turned in the autumn, when American and French forces succeeded in trapping a British army at Yorktown.

(Yorktown Battlefield Marker)

YORKTOWN BATTLEFIELD
YORKTOWN, VIRGINIA

The Yorktown Battlefield is maintained by the National Park Service as a Colonial National Historical

> The American Revolutionary War affected all aspects of American Society. Patriotic men joined militia units or the Continental Army; others, loyal to Great Britain entered into provincial service. Some women and children followed their husbands and fathers, while others were left behind to manage the family farm or business. Enslaved African Americans also had opportunities to participate on either side of the war, and in some cases were offered freedom for their service. At war's end, several thousand African Americans had served as soldiers, spies, guides, sailors, ship pilots, and laborers.
>
> The first chance to serve for freedom was for the British. On November 7, 1775, Virginia's royal governor, John Murray, Earl of Dunmore, in an attempt to raise forces, proclaimed "all indentured servants, Negroes, or others (appertaining to rebels) free, that are able and willing to bear arms in the King's cause." By December, nearly 300 African Americans in uniform, with the words "Liberty to Slaves" inscribed across their breasts, were members of "Lord Dunmore's Ethiopian Regiment." A few days later, on December 9, at the Battle of Great Bridge, a British force of 600, nearly half of which was African American, was thrown back by the Patriots. By August 1776, Dunmore was forced to abandon Virginia.
>
> (Yorktown Battlefield Marker)

Park. There is a small park entrance fee that is payable at the visitor center where the events of the siege of Yorktown are presented in a theater program and several exhibits.

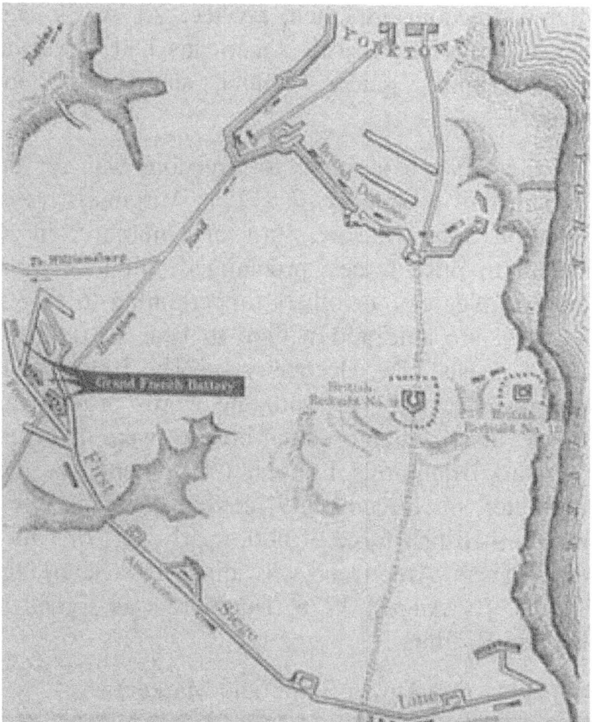

The earthworks in front of you and across the open plain played an integral part in the battle that won America's independence. For many American, French, British, and German soldiers, only these fortifications stood between them and the enemy during the 1781 Siege of Yorktown.

Please help the National Park Service preserve these works as monuments to soldiers who fought and died here. Walk only on designated trails. To protect the earthworks, bicycling, sledding and other similar activities are prohibited.

"Persons whose duty does not call them to the trenches and who assemble there merely to indulge curiosity are to walk on the reverse of the trenches that they may not interrupt the works."

General George Washington

General Orders, Headquarters before York, Tuesday,

October 9, 1781

(Yorktown Battlefield Marker)

From the visitor center, there is a seven-mile battlefield tour with six major sites that chronologically follow the events of the siege. Portions of the tour road are on busy thoroughfares. Caution is recommended because local drivers speed through these areas in contrast to the slow moving tourists.

Site A. British Inner Defense Line — Early in August 1781, Cornwallis began to fortify the Town of York and Gloucester Point across the river. On September 28th, the Allied Army of American and French forces, numbering more than 17,000 men, marched to Yorktown for the siege of Cornwallis' garrison of about 8,000 men. After their arrival, Cornwallis withdrew his troops from most of his outer defenses to consolidate his position behind the earthworks of his inner defense line.

The Inner Line

Here this tour road uses a gap through the old British defenses. The works you see today are an enlargement by Confederate troops who re-fortified Yorktown in 1862. Just ahead, shot and shell were heavy in the fateful days of October 9-17, 1781.

(Yorktown Battlefield Marker)

Site B. Grand French Battery — After a week of preparation, the Allied Army of American and French forces under the cover of darkness constructed its first siege line during the night of October 6th, 1781. The

The First Siege Line (reconstructed)

Washington's troops built works like these when they opened the Siege of Yorktown. These fortifications, together with those of the French, made a circling line that stretched a mile and a quarter, from the York River to Yorktown Creek.

(Yorktown Battlefield Marker)

siege line extended from the Grand French Battery to the York River. On October 9th, Allied artillery opened fire on the British, and the bombardment began. The Grand French Battery was the largest gun emplacement on the first siege line.

First Allied Siege Line

The Grand French Battery

The Grand French Battery formed part of the first Allied siege line around Yorktown. French and American soldiers worked through the night of October 6, 1781, to establish the line from a point about one hundred yards across the road to the York River.

Largest gun emplacement built during the siege, the Grand French Battery opened fire on October 10 and inflicted great damage on the British. A French officer noted that "... the superiority of the fire of these different batteries over those of the enemy, silenced the firing from the town." This superiority in artillery contributed greatly to Allied victory.

The National Park Service has reconstructed most of the Grand French Battery on its original site and equipped it with original and replica Revolutionary War period artillery.

(Yorktown Battlefield Marker)

In the 1930's, portions of the American and French lines were reconstructed using the same methods employed by General Washington's army.

(Yorktown Battlefield Marker)

Grand French Battery

These three batteries and redoubt were a continuation of the Grand French Battery complex across the road. Occupying a section 1,000 feet from east to west, this battery complex contained approximately thirty pieces of artillery which bombarded the main British Defense a half-mile away. So effective was its work that the digging of the Second Parallel could take place one day after the Grand French Battery opened fire.

(Yorktown Battlefield Marker)

Site C. Second Allied Siege Line — On October 11th, Allied troops began a second siege line within point blank artillery range of the British. However, the line could not be completed because two small, detached British earthen forts, Redoubts 9 and 10, blocked the way to the river.

Artillery and the Siege

Cannon, howitzers and mortars in concentrations, such as you see here, counted heavily in the allied success at Yorktown. The allies confronted Cornwallis with superior "firepower" and numbers. For this the British had no answer but surrender.

(Yorktown Battlefield Marker)

Second Allied Siege Line

Sieges—18th century style. Eighteenth century military commanders conducted sieges according to formal step-by-step procedures perfected over centuries of European warfare. The siege of Yorktown followed the principle of digging successive trenches ever closer to the enemy fortifications. This tactic was still being used in World War I.

(Yorktown Battlefield Marker)

Site D. Redoubts 9 and 10 — On the night of October 14th, French troops attacked Redoubt 9 while American troops stormed Redoubt 10, capturing both positions in less than 30 minutes. This allowed the Allies to complete the second siege line and

Redoubts 9 and 10

"The ... batteries were principally directed against ... the enemy's advanced redoubts on their extreme ... left to prepare them for the intended assault ..." General George Washington's Diary, October 14, 1781.

The completion of the Allied Second Siege Line was blocked by a portion of the British outer works — two detached earthen forts called Redoubts 9 and 10, located 400 yards in advance of the British Inner Defense Line. Though General Washington considered allied frontal attacks against the British Inner Defense Line impractical and costly, Redoubts 9 and 10 were a different matter. General Washington called upon the infantry to capture these positions.

On October 14, Allied artillery bombarded Redoubts 9 and 10 most of the day, preparing for American and French assaults. At approximately eight o'clock in the evening, the artillery briefly fell silent. Then several cannons fired in unison, the signal for the attack columns to move on their objectives...

"Last evening the enemy carried my two advanced redoubts on the left by storm, and during the night have included them in their second parallel ... my situation now becomes very critical ..."

General Charles Lord Cornwallis to General Sir Henry Clinton, October 15, 1781.

(Yorktown Battlefield Marker)

Redoubt 9

"Anyone can imagine what happened once we were inside the redoubt. People of four nations were thrown together: Frenchmen, English, Scots, and Germans … the soldiers … were so furious that our people were killing one another: The French were striking down everyone in a blue coat. Since the Deux-Ponts wore blue, many of us were stabbed to death. Some of the Hessian and Anspach troops [German units in the British army] wore uniforms almost identical to ours, and the English wore red that in the dark of night seemed blue as well, so things went very unmercifully…" Private Georg Daniel Flohr, Royal Deux-Ponts Regiment.

Baron Antoine de Viomenil had overall command of the French attack on Redoubt 9 which was defended by approximately 120 British and German forces. Viomenil selected Colonel en Second Guillaume des Deux-Ponts to head the 400-man assault column from the Gatenais unit and the German recruited Royal Deux-Ponts Regiment.

The French approach was delayed as pioneers (engineering troops) cleared obstructions protecting the British perimeter. Darkness, the chaos of close-quarter combat and the confusion of German commands coming from both attackers and defenders compounded the difficulties of the assault. Within 30 minutes however, Viomenil was congratulating Deux-Ponts on taking the redoubt.

The last obstruction to the Allies completing their Second Siege Line had fallen.

(Yorktown Battlefield Marker)

On this line at the Siege of Yorktown on October 14, 1781, at night the battalions of Gimat, Hamilton and Laurens of the light Infantry Division of Major-General Marquis De Lafayette under the command of Lieutenant-

Colonel Alexander Hamilton with unloaded muskets and fixed bayonets, scaled the parapets and gallantly captured Redoubt No. 10 which formed the extreme left outlying defense of the British and hastened the surrender of Yorktown in the last fighting of the troops of General George Washington.

Erected by the Sons of the Revolution in the State of New York, 1925.

(Yorktown Battlefield Marker)

Redoubt 10

"The column marched in silence, with guns unloaded, and in good order. Many, no doubt, thinking that less than one quarter of a mile would finish the journey of life..." Captain Stephen Olney, 1st Rhode Island Regiment.

The Marquis de Lafayette selected the light infantry companies from his Continental Army division to carry out the American attack on Redoubt 10. Lieutenant-Colonel Alexander Hamilton led the assaulting force of 400 Americans against the 70 British defenders.

(Yorktown Battlefield Marker)

construct a Grand American Battery between the two redoubts.

Cornwallis, realizing his forces were trapped, attempted to move them across the York River and escape on the night of the 16th. However, the attempt was abandoned when a storm interrupted the crossing and prevented its completion.

Concerned an accidental musket firing would give away their advance, the light infantry approached the redoubt with unloaded weapons. The British discovered the attack column as it neared the redoubt and vigorously defended their position. The Americans surged through the abatis obstructions surrounding the redoubt, then through holes in the palisaded log walls within the trench, breached the top redoubt, and within 10 minutes of mostly hand-to-hand combat, secured their objective.

Alexander Hamilton, in his report to Lafayette on the action, noted that the British were "intitled to the acknowledgment of an honorable defence."

(Yorktown Battlefield Marker)

Time, Tide and Erosion

On your left is British Redoubt No. 10 partly reconstructed where a fragment of its moat was found in 1956. The remainder of it, as well as parts of adjacent works, was washed to sea during the 175 years of crumbling riverbanks.

(Yorktown Battlefield Marker)

Grand American Battery

"The Works which we have carried are of vast importance to us. From them we shall enfilade the enemy's whole line..." General George Washington to the President of Congress, October 16, 1781.

The capture of British Redoubts 9 and 10 enabled the Americans to quickly finish the Allied Second Siege Line, constructing this Grand American Battery within point blank artillery range of the British Inner Defense Line. On October 17, Continental artillery crews began bombarding the British from their new battery. According to Lieutenant-Colonel Henry Dearborn of the 1st New Hampshire Regiment, the allied barrage that day "made the enemies situation so very disagreeable — that about the middle of the day his Lordship was induced to send out a flag, with some proposals for a capitulation."

The following day officers from both sides met at the home of Augustine Moore to negotiate the British surrender terms.

(Yorktown Battlefield Marker)

The second siege line was completed the morning of October 17th. That same day, after nine days of round-the-clock bombardment and now facing direct fire from Allied artillery, Cornwallis requested a cease-fire to discuss surrender terms.

The Second Siege Line

The low earthworks on either side are part of the French and American advance line. Construction was started when allied artillery fire had sufficiently weakened the British defenses. The line was completed after the capture of Redoubts No. 9 and 10.

(Yorktown Battlefield Marker)

A British Sortie

Before daybreak on October 16, across the field to your right, some 350 British soldiers attacked the allied inner line near the point were the French and American sectors joined. The attack created alarm but as Washington relates, it was "small and ineffectual."

(Yorktown Battlefield Marker)

In the final major campaign of the war, the British continued to use Africans. Moving into Virginia in May 1781, General Charles Lord Cornwallis and other British commanders pressed African Americans into labor. An African American corps built fortifications at Portsmouth and approximately 2,000 were employed to construct earthen defenses at Yorktown and Gloucester Point.

Trapped in Yorktown by General George Washington's army, Cornwallis' forces labored under bombardment, confinement and dwindling supplies. Smallpox broke out, and the African Americans were very susceptible to the disease. In what some perceived as an attempt to inflict the smallpox on the enemy, the African Americans were forced out of the British defenses towards the American lines.

(Yorktown Battlefield Marker)

U.S. National Cemetery Established 1866

For twenty-nine days in the spring of 1862, Yorktown was visited by another siege. In this second year of the Civil War, a Union army driving on Richmond was temporarily stalled by a thin Confederate line. Dead of both sides lie here.

(Yorktown Battlefield Marker)

Communicating Trench

To your left is a rebuilt trench dug to protect advancing troops from shellfire — heavy in this open area. The trench followed a zigzag pattern so as to keep a bank of earth between the men in the trenches and British cannonading.

(Yorktown Battlefield Marker)

Site E. Moore House — On October 18th, officers from both sides met at the home of Augustine Moore to negotiate the surrender terms for Cornwallis' army.

Moore House

The battlefield was quiet October 18, 1781 under temporary truce. That day two Allied commissioners met with two British [commissioners] and drafted the terms of surrender. They met in this house, now restored and refurnished. It was then the home of Augustine Moore, merchant of Yorktown.

(Yorktown Battlefield Marker)

Wormley Creek Crossing

The earth dam in front of you creates Wormley Pond which existed in 1781 when Augustine Moore operated a gristmill here. American troops marched over the dam regularly as they moved to and from the Siege line.

(Yorktown Battlefield Marker)

The Deposit

Eighteenth century siege warfare meant weeks of labor for the soldiers, not only on the trenches, but in the rear areas where special siege materials were made and stored.

In constructing the siege lines, engineers and sappers used quantities of gabions (wickerwork-like baskets), fascines and saucissons (bundles of stout wooden sticks), and fraises (heavy pointed poles). These materials of war were temporarily stored in this ravine until American troops were ready for them on the line.

(Yorktown Battlefield Marker)

Site F. Surrender Field — On October 19th, Cornwallis' army marched onto Surrender Field and laid down its arms. However, missing from the surrender formalities was the British Commander, Lord

Surrender Road

The course of the road here is about as it was in colonial times. On October 19, 1781, in the afternoon, the solders of Cornwallis' Army marched this way and filed off into the field on your left. This was the place designated for the formal capitulation of the British Garrison.

(Yorktown Battlefield Marker)

Cornwallis, who claimed he was indisposed. Instead, his deputy, General O'Hara attempted to surrender his sword to General Rochambeau. However, Rochambeau, recognizing a possible attempt to not recognize the American forces in the victory, refused to accept the sword and indicated that it should be given to General Washington. However, Washington, unwilling to accept the sword from a deputy, also refused the sword and indicated that it should be surrendered to his deputy, General Lincoln. Lincoln accepted the surrender sword.

When the British Prime Minister, Lord North, learned of Cornwallis' defeat, he reportedly said, "Oh God! It is all over!" The American victory at Yorktown, the last major battle of the American Revolution, secured independence for the United States and significantly changed the course of world history.

After the British surrender, the French forces remained in the Yorktown area during the winter while Washington and most of the American troops returned to New York to keep General Clinton bottled-up in New York City. In the summer of 1782, General Rochambeau and his troops departed for New England and left for France from Newport, RI, in December. Washington kept the American army intact for several more months, until the Treaty of Paris officially ended hostilities between the British and Americans in September 1783.

The Surrender of Cornwallis at Yorktown

Print, Ca. 1781, Paris, Hand-colored line engraving.

(Yorktown Visitor Center Exhibit)

While the war's end brought freedom to some African Americans serving under the British, there were also opportunities on the Patriot side for African Americans to gain freedom. Some served individually, alongside owners, or in place of them and were granted manumission after the war. Others became emancipated upon enlistment into military service as was the case with the First Rhode Island Regiment, which participated in the Siege at Yorktown.

(Yorktown Battlefield Marker)

American and French Encampments — The Yorktown National Historic Park also offers another driving tour deeper into the countryside. The tour includes the American and French encampment areas as well as the site of Washington's headquarters.

American Artillery Park

At a good vantage point beyond the road to your left, "Chief of Artillery" Henry Knox laid out the station where American ordnance was assembled, serviced and, as required, repaired. The French maintained a similar area in their own encampment.

(Yorktown Encampment Marker)

Washington's Headquarters

General George Washington established his headquarters in this area at the junction of the American encampments to the east and the French encampments to the north.

Here he set up two tents: a large one for meeting with his staff and for dining, and a smaller one as his private office and sleeping quarters. The house nearby may have served as additional shelter, while the spring over the brow of the hill to the right offered fresh water.

Washington is represented with other members of his staff, planning the attack on British Redoubts Numbers 9 and 10.

(Yorktown Encampment Marker)

The Town of York — A walking tour of downtown Yorktown can be easily made from the visitor center. There is a walkway and bridge to the town where several sites can be visited.

"York-Town, capital of the county of that name, is situated on a rising ground, gently descending every way into a valley, and tho' but stragglingly built, yet makes no inconsiderable figure." Edward Kimber, observations in several voyages and travels in America in the year 1736.

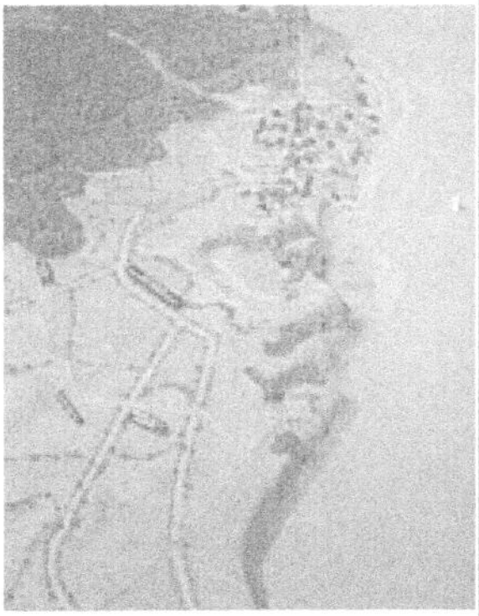

Before the American Revolutionary War, Yorktown was a prosperous, thriving tobacco port. Approximately 80% of the town was damaged or destroyed during the climactic military campaign of that war, General George Washington's victory over British General Charles Lord Cornwallis in the 1781 siege.

While the town never regained its colonial prominence, Yorktown is still a vibrant community, boasting shops, restaurants, places of lodging, churches and county courthouse. Reminders of the town's 18th century history, such as historic homes and buildings, and military earthworks, can yet be found providing visible symbols of Yorktown's important contributions to the founding of the United States of America.

"The British pull'd down many of the houses in York and those they left have been much injured by the bombardment..." William Reynolds to William Murdoch, October 26, 1781.

(Yorktown Marker)

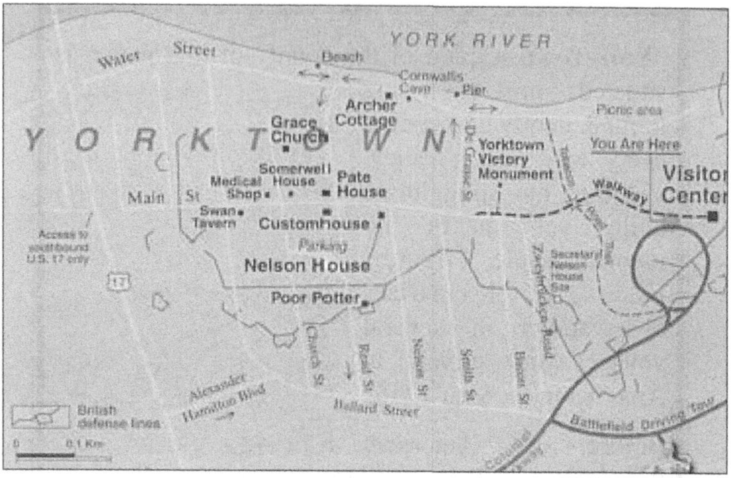

Tobacco

Cornerstone of Yorktown Commerce

English colonists found no Aztec gold or Inca silver, but they did find wealth in another American Indian product — tobacco. By 1700 the tobacco habit had spread over much of the world and created a boundless market for fine Virginia leaf. A whole commerce grew based on tobacco, with plantations, warehousing, shipping, and official inspectors to assure high quality.

(Yorktown Battlefield Marker)

"This Yorktown, or Little-York, is a small city of approximately 300 houses: it has, moreover, considerable circumference. It is located on the bank of the York River, somewhat high on a sandy but level ground. It has 3 churches, 2 Reformed English and 1 German Lutheran … and 2 Quaker meeting houses, and a beautiful court or meeting house…" — Johann Conrad Doehia, Anspach-Bayreuth Battalion, British army, July 31, 1781.

Travel a short distance along this walkway to discover historic Yorktown.

Established in 1691, by the 1750s the town was a thriving seaport supporting a prosperous waterfront business district with wharves, warehouses and pubs. Stately and moderate homes, outbuildings, stores, churches, taverns and government buildings filled the bluff overlooking the river.

The town was extensively damaged during the 1781 siege and never recovered its colonial prominence.

Explore Yorktown and discover among a still active community sights that reflect the town's past glory: earthworks encircling the town, the 1881 Monument to the Alliance and Victory, and 18th century buildings, including the home of a signer of the Declaration of Independence.

"I have this day visited the town of York, to witness the destructive effects of the siege. It contains about sixty houses…many of them are greatly damaged and some totally ruined, being shot through in a thousand places and honey-combed ready to crumble to pieces." — Dr. James Thacher, Surgeon, Continental Army, October 22, 1781.

(Yorktown Marker)

Secretary Nelson House — The home was Cornwallis' headquarters during the siege of Yorktown.

Foundations of the home of Thomas Nelson, president of the council and secretary of state of the colony of Virginia erected in 1725. First headquarters of Lord Cornwallis it was destroyed during the siege of Yorktown in 1781.

Preserved and marked by the Yorktown Branch Association for the Preservation of Virginia Antiquities, 1930.

(Yorktown Marker)

Secretary Nelson House

Thomas, a son of "Scotch Tom," who established the Nelson family in Yorktown, was for many years Secretary for the Colony of Virginia. This comfortable and spacious home attracted Lord Cornwallis who needed a headquarters near his main defense on the east side of town. The British commander remained here through much of the siege — until Allied gunners made the house untenable. Its foundations are to your left.

(Yorktown Marker)

A l'occasion du bicentenaire des traites de Paris et de Versailles qui consacrerent de 3 Septembre 1783 l'independence des Estats—Unis d'Amerique, cette plaque a ete apposee par les Filles de la Revolution Americaine en temoignage de gratitude envers les artisans qui s'etaient vous "au plus beau des quvrages — delaboration de la paix" (Benjamin Franklin)

On September 3, 1983, the bicentennial of the treaties of Paris and Versailles which secured the peace and established the independence of the United States of America, this tablet was offered by the National Society Daughters of the American Revolution as a grateful tribute to the architects of these treaties who dedicated themselves to "the best of all works — the work of peace." (Benjamin Franklin)

(Yorktown Marker)

Dudley Digges House, circa 1760

"...Lieutenant-colonel Tarleton directed them to charge into the town, [Charlottesville, Virginia] ... and to apprehend, if possible, the governor and assembly. Seven members of the assembly were secured ... and several officers and men, were killed, wounded, or taken." Lieutenant-Colonel Banastre Tarleton, A History of the Campaigns of 1780 and 1781.

Dudley Digges built this classic Virginia Tidewater style home around 1760. The outbuildings, wellhouse, kitchen, granary, and smokehouse are typical of those found in the colonial era. The house was restored in 1960 and the outbuildings were reconstructed by the National Park Service in the 1970s.

One of the members of Virginia assembly captured by the British during their Charlottesville raid on June 5, 1781 was the former lieutenant governor, Dudley Digges ... Throughout the war, Dudley remained active in numerous areas of Virginia government, including helping to write the commonwealth's first constitution and becoming one of the first members of the state council.

Dudley's home, like so many other Yorktown houses, was damaged during the 1781 siege and rendered uninhabitable. Dudley moved to Williamsburg and died there in 1790.

(Yorktown Marker)

Yorktown's prosperity as a port peaked around the middle of the 18th century. By 1781, Main Street, as depicted in this painting, had become a quieter, less active avenue.

Painting by Sidney King

(Yorktown Marker)

Dudley Digges House — Built about 1760 by Digges, a lawyer and Virginia government official, 1752-81. Along with Daniel Boone, Digges was also one of the seven members of the Virginia assembly that was captured during the surprise British raid on Charlottesville (a previous site on this road trip).

Nelson House — It was the home of Thomas Nelson Jr., who led the local Yorktown "tea party" and tossed the tea that was consigned to the Prentis Store in Williamsburg off a merchant ship in Yorktown harbor in November 1774. From 1761 to 1775, Nelson served in the Virginia House of Burgesses and then two years as a delegate to the Second Continental Congress in Philadelphia where he signed the Declaration of Independence. In June 1781, he

Nelson House

"Scotch Tom" Nelson, founder of the Nelson family and fortune in Yorktown, built this mansion in the early 1700's. Later it passed to Yorktown's most illustrious son Thomas Nelson, Jr., who signed the Declaration of Independence, served as wartime Governor of Virginia, and commanded the Virginia militia during the siege of Yorktown.

(Yorktown Marker)

Grace Church

The sturdy marl walls of this church have been standing since 1697 despite the ravages of war during the two sieges of Yorktown (1781 and 1862) and despite the "great fire" of 1814. It continues, as it has for generations, to serve York-Hampton Parish in the Protestant Episcopal Diocese of Southern Virginia.

(Yorktown Marker)

was elected the third governor of Virginia, succeeding Thomas Jefferson, and, with the rank of brigadier general, commanded the Virginia militia at the siege of Yorktown.

Grace Church — Built about 1697, this still-active church served the York-Hampton Parish in colonial times.

Swan Tavern — This reconstructed tavern is built on the original 1722 site.

The Swan Tavern

For 133 years, the Swan Tavern was a leading place of public entertainment in the busy seaport "Town of York." Built before 1722, it was destroyed when a nearby ammunition magazine exploded in Civil War days. Tavern, kitchen, smokehouse, stable and dairy have been reconstructed.

(Yorktown Marker)

Medical Shop (Reconstructed)

Dr. Corbin Griffin was a prominent Yorktown physician active in the American Revolutionary War, serving as a surgeon with Virginia forces. During the Siege of Yorktown, he was imprisoned by the British on a ship anchored in the York River. During his confinement, his "cellar down the water side" was pilfered by sailors from the British sloop Bonneta while other British troops took items from his medical shop.

"I must ... request that your Lordship will inform me of the Reason of Dr. Griffin's confinement on Board of one of your Prison Ships." — Governor Thomas Nelson, Jr. to General Charles Lord Cornwallis, September 25, 1781.

(Yorktown Marker)

Medical Shop — This is the reconstructed 18th-century medical shop of Dr. Corbin Griffin. Dr. Griffin was imprisoned by the British during the siege of Yorktown.

Victory Monument — Authorized by the Continental Congress after the Allied victory in 1781, construction began a hundred years later during the centennial celebration of the surrender in 1881. The monument, which is nearly a hundred feet high, is topped with a figure of Liberty.

Yorktown Victory Monument

This monument was authorized by Continental Congress, October 29, 1781, just after the news of surrender reached Philadelphia. Actual construction began 100 years later and was completed in 1884. The original figure of Liberty atop the Victory shaft was severely damaged by lightning. A new work replaced it in 1956. The shaft of Maine granite is 84 feet in height to which Liberty adds another 14 feet.

(Yorktown Marker)

Overnight in the Yorktown-Williamsburg Area — There are many possibilities for lodging in the area, but the Colonial Houses of Williamsburg present a unique opportunity to stay in an 18th century, colonial home in the heart of Williamsburg. Each home is furnished with period reproductions and range in size from small, single bedroom homes to large family homes with over a dozen rooms. A home can be reserved by calling 1-800-447-8679 or on the world wide web at www.colonialwilliamsburg.com.

In memory of the men of the French fleet who died in the naval phase of the Yorktown campaign in the battle off the Virginia Capes on 5 September 1781.

In appreciation of the service of Admiral Francois Joseph Paul de Grasse who brought his entire West Indies fleet to the aid of the Americans, who transported more than 3,000 French troops under the command of the Marquis de Saint-Simon to reinforce the land forces and who commanded the French fleet in the battle off the Virginia Capes, defeating a squadron of the British Royal Navy under Admiral Thomas Graves. General George Washington called French naval superiority "the pivot upon which everything turned" as the French victory prevented General Earl Cornwallis from aid or escape by sea and allowed a smaller French fleet from Newport under Commodore de Barras to enter the Chesapeake Bay with siege guns essential in the land battle that followed.

In commemoration of the Bicentennial of this most important naval battle fought in American waters which assured the victory at Yorktown and American Independence.

This tablet is erected by the National Society of the American Revolution, October 18, 1981.

(Yorktown Marker)

"We are now friends with England and with all mankind." Benjamin Franklin, 1783, American Peace Commissioner.

"This great peace monument is a symbol of the sacrifices in lives and property in the Revolutionary War, which ended at Yorktown and which brought us our independence. It symbolizes, too, the peace between the mother country and America — a peace not seriously interrupted since 1781," Horace M. Albright, 1931, Director, National Park Service.

"The Treaty of Paris was the first step toward an alliance with Great Britain which has grown stronger through two centuries to become one of our most important alliance relationships. Political, cultural, economic, and defense ties between our two nations are firm and lasting," Ronald Reagan, 1983, President, United States of America.

Dedicated 19 October 1984, Bicentennial Year, Ratification Treaty of Paris.

(Yorktown Marker)

At Yorktown, October 19, 1781, after a siege of nineteen days by 5,500 American and 7,000 French troops of the line, 3,500 Virginia Militia under command of General Thomas Nelson and 35 French ships of war, Earl Cornwallis commander of the British forces at York and Gloucester surrendered his army of 7,251 officers and men, 840 seamen, 244 cannon and 24 standards to his excellency George Washington, Commander in Chief of the combined forces of America and France to his Excellency the Comte de Rochambeau commanding the auxiliary troops of his most Christian Majesty in American and to his Excellency the Comte de Grasse, Commanding in Chief the Naval Army of France in Chesapeake.

(Yorktown Marker)

BIBLIOGRAPHY

Bioguide.congress.gov, "Giles, William Branch, 1762-1830," 2004.

Bobrick, Benson, **Angel in the Whirlwind,** Penguin Books, New York, 1997.

Colonial Williamsburg Foundation, **Official Guide to Colonial Williamsburg**, 1998.

Colonial National Historical Park, Virginia, National Park Service, U.S. Department of the Interior, "Yorktown," GPO: 454-767/00136, 1999.

Colonial National Historical Park, Virginia, National Park Service, U.S. Department of the Interior, "Colonial," GPO: 454-767/60433, 1999.

Fodor's Travel Publications, **The Thirteen Colonies**, 2003.

Geocities.com/stpeterstc/, "St. Peter's Parish Church, New Kent County, Virginia," 2004.

Metmuseum.org, "George Washington Crossing the Delaware," 2004.

Myerchin.org, "General Friedrich Adolph Riedesel," 2004.

Odur.let.rug.nl, "John Marshall (1755-1835)," 2004

Powhatanva.com, "Welcome to Powhatan, Virginia," 2004.

Patrick Henry Memorial Foundation, "Red Hill, Patrick Henry National Memorial," 2003.

Purcell, L. Edward and David F. Burg, Editors, **The World Almanac of the American Revolution**, Pharos Books, New York, 1992.

Richmond Metropolitan Convention & Visitors Bureau, "The Story of Virginia," 2002.

Scottsville.org, "Scottsville on the James," 2004.

Shively, Julie, **Places of the American Revolution,** Ideals Publications, 2001.

St. John's Church, "St. John's Church, Richmond, Virginia," 2004.

Symonds, Craig L., **A Battlefield Atlas of the American Revolution**, Nautical & Aviation Publishing Company of America, 1986.

Thomas Jefferson Foundation, Inc., "Monticello," 2002.

Thomas Jefferson Foundation, Inc., "Mulberry Row," 2002.

Raymond C. Houghton is a retired college professor, former Federal Government staff member, Vietnam Veteran and one-time, General Electric employee. He has honors from the Department of Commerce, is listed in Who's Who in America, holds a doctorate from Duke University and is the historian for the Town of Bethlehem, NY. His goal is to get people out on our historic highways to relive and appreciate an important time in American history.

Made in United States
North Haven, CT
10 November 2026

22088270R00088